A
Man's
Tall
Dream

THE STORY OF EASTWOODHILL

A
Man's
Tall
Dream

THE STORY OF EASTWOODHILL

JOHN BERRY

Foreword by
DAVID BELLAMY

First Published 1997 by
The Eastwoodhill Trust Board
PO Box 146 Gisborne, New Zealand

ISBN No. 0-473-04561-3

Typeset in Palatino
Design, Scans, Imagesetting and Printing by
Te Rau Herald Print
PO Box 945 Gisborne, New Zealand

For Tess, Emma and Abby,
with the prayer that they grow up
in a green and friendly world.

Acknowledgements

The Eastwoodhill Trust Board gratefully acknowledges the generous assistance towards the publication of this book from:

 The Founders' Society of New Zealand

The C. ALMA BAKER Trust

The Survey and Land Information Department.

Author's Acknowledgements

It would be foolish, for fear of omission, to attempt to acknowledge everyone who assisted with information, or in other ways, in compiling this book. Special thanks, however, are due to Garry Clapperton, for the use of invaluable material from his own research; to Ivan Mitchell, for data contained in his "Sambucus" writings in the Gisborne Herald; to Graham Weston, for making available private correspondence from Douglas Cook; to Professor David Bellamy, for his Foreword; to H.B. (Bill) Williams and the Book Committee (Rodney Faulkner, Bett Chrisp, Terence Williams and Paul Pollock); to Lee Newman, secretary of the Eastwoodhill Trust Board; to Sheridan Gundry, for her editing skills; to Joyce Sherriff, for assisting in most difficult circumstances; to Dot Cassin for continuing encouragement; and to my wife Bette, for her support.

Edited by
Sheridan Gundry, GEMS Communications

Historic black and white photograph selection, captions and list of genera by Eastwoodhill curator Garry Clapperton.

All unmarked photographs, including many by W. D. Cook, were sourced from the Eastwoodhill archives.

Contents

WILLIAM DOUGLAS COOK

Foreword

By Professor David J. Bellamy

I have travelled the world for much of my life and have been to many magical places. Eastwoodhill is very special for it fills me with hope, thanks to the inspiration of the power of plants and people working together in symbiosis. The power to put the world back into sustainable working order.

I had read a lot about Eastwoodhill, I had even seen pictures of it, but on February 10, 1989, I saw it for myself for the first time. What a shock, no not of disappointment, for it lived up to all my expectations, but one of amazement at how the vision of one individual had created such an oasis of hope in such a blasted landscape.

The hills all about were in tatters, the scars of massive erosion glaring white pumice in the hot summer sun. Cyclone Bola had done her worst the year before, wreaking havoc on the overgrazed uplands where the stormproof native bush had been all but eliminated. Eastwoodhill Arboretum had held firm, a green oasis at the heart of Poverty Bay.

Fifty years before, the torrential rain associated with another cyclone had done its worst, ripping the landscape apart and creating the infamous Tarndale Slip. Yet even then the trees of Eastwoodhill had held firm, trees drawn from the four corners of the Earth thanks to the vision of William Douglas Cook.

David Bellamy expressively addresses the crowd gathered in the Daffodil Patch in February 1989, flanked by Friends chairman Nisbet Smith, board member Terence Williams, with H.B. Williams obscured by David Bellamy. (Photo Friends of Eastwoodhill)

Douglas had returned from a Europe torn by the depredation of World War 1. He was worried that the great estates of Europe would be destroyed, along with their genetic stock of trees, gleaned from all around the world.

So on land he had purchased in the rolling hill country of the Ngatapa district near Gisborne, he started to create his oasis; a Garden of Eden held altogether by the best of the world's trees he had seen and heard about.

Douglas Cook worked with his garden until he died in 1967, importing the best genetic stock he could from nurserymen and botanic gardens the world over. He and his gardeners potted, planted and planned. When drought threatened their tree nurseries, they carried water by the bucket to wean seedlings into saplings and saplings into truly magnificent trees. Indeed, the rates of growth of trees as diverse as English oak and Chinese maples, ash and liquidamber have continued to amaze visitors. "Hardwood trees can't grow that fast" the sceptics said, until they came and believed their astounded eyes. It is a truly wonderful place, ranking alongside the great arboreta of the world.

A great, green genetic bank for all our futures.

1   A Fledgling Farmer

Ｈow does a dream begin? Perhaps the tall dream of William Douglas Cook was already forming in his mind when, as a 25-year-old fledgling farmer on March 31, 1910, he looked across his scrubby, uninviting land amid the inland hills of Ngatapa on the eastern coast of the North Island, New Zealand. It must have been a barren scene, even a daunting one, to a young man who had always had an eye for beauty. Many years later he was to write that in his secret heart he had envisaged his first cow paddock of six acres one day becoming a garden and shrubbery – "I dared not tell a soul of my hope as it seemed, in those days of struggle, almost stupid to cherish such an ambitious hope." His dream, like the trees he planted, was to grow taller with the years. It developed a grander dimension when, convalescing in Britain from wounds he received in World War 1, he visited the grand homes and gardens of relatives and their friends.

In May 1910, six weeks after occupation, Douglas Cook's new cottage is sited 150 feet above the Burnside flat. The Taumatapoupou Stream was the only water source on the property and in summer dried up. Water was pumped to tanks with a gasoline-powered pump. The sandstone seam in the stream was the crossing point of the historic footpath of the Te Aitanga-a-Mahaki people to Ngatapa pa.

In 1910 the slope leading up to the cottage was barren whereas today the woolshed and yards in Pear Park are nestled in trees. The telephone line crossing the scene was installed by J.B. Broadhurst in late 1910 at a cost to Douglas Cook of £12/10/-. Compare this scene with a 1950s aerial view of the same area in Chapter 8.

Sheep work in the yards in 1911. The photograph shows the cottage (at left), Orchard Hill (centre) and Arateitei (right), the high path used by the Maoris on their way to Ngatapa.

On that March day in 1910 he had experienced an unceremonious arrival in the Gisborne district. He had travelled north by sea from Hawke's Bay and the ship lay out in Poverty Bay (named by Captain James Cook in 1769 as his inhospitable first landing place in New Zealand). Douglas Cook was landed in a wicker basket, swung out from the ship's derrick, on to the deck of a bobbing tender far below. He later recalled: "Everything in the district was in keeping. It was backward because of its isolation. All communication was by sea. There were bush tracks communicating with the outside world but there was no such thing as an all-weather road either north or south till after World War 1. One could drive a buggy south in fine weather in summer, but there was no road north." He travelled by bullock wagon to Ngatapa to see his land for the first time.

Douglas Cook had been growing peaches in Hastings before he acquired his property at Ngatapa, 35km by road north-west of the coastal town of Gisborne, by a successful draw in a land ballot. It had been Maori land and was now part of "the Ngatapa Settlement". In pre-European times Maori grew taro and kumara in the warmer Waikakariki valley up as far as the Ngatapa School. The forested and scrub-covered hill country was used for seasonal bird gathering of weka and native pigeon. There are earthworks as evidence of that early Maori presence. There was a pathway marked on an early map as "the old track to Ngatapa". It was the route to Ngatapa pa from Repongaere, a settlement near Patutahi. Ngatapa was a place of refuge in times of trouble and the path also gave access to the inland Urewera Country.

William Scott Greene was the first European to run sheep in the area in 1870 and took up a lease of what was called the Okahuatiu block from the Maori in 1873. In 1875 members of Te Whanau-a-Kai, a subtribe of the Te Aitanga-a-Mahaki, were awarded ownership of a 25,268 acre area named Okahuatiu No.1 Block, and from those owners Scottish cousins John Clark and David Dobbie leased the run in 1877. Their partnership was dissolved in 1882 after which James Williamson, noted Auckland financier and a partner in the Auckland Agricultural Company, and a Mr Rose, purchased it and called it Okahuatiu Station. Dobbie stayed as manager until 1899. The block was passed to Williamson's three sons who later subdivided it between themselves for taxation purposes. The blocks were named Ngatapa, Hihiroroa and Taumata.

Douglas Cook wrote of his newly-acquired land: "When I came to this property in 1910 it was practically virgin land, some in English grasses, some in native grasses, but most of it still covered with manuka scrub which is botanically known as *Leptospermum*."

Writing in the Journal of the Royal Horticultural Society in 1949 he was to observe that no garden in New Zealand at that time was more than 100 years old and that, in the early days of the first settlers, few "really good things" were planted. Sailing ships taking several months for the voyage from Britain were not the best means of transporting live trees, the first steamer to New Zealand being a single voyage of the Stad Haarlem in 1879, and no regular steam service till the New Zealand Shipping Company's steamers began in 1883. Nevertheless, he wrote, by the 1880s many fine gardens had been established and large public parks laid out.

In the matter of early public parks in New Zealand, it is interesting that, at the time of Douglas Cook's birth in New Plymouth in 1884, a board of trustees in that town, set up under the Taranaki Botanic Gardens Act 1875, was in the early stages

Douglas Cook's original of the Miller family crest was lost when his home burned down in January 1931. This copy was drawn for him by Ernest Hindman, later a brother-in-law to Bill Crooks, who then worked for neighbouring farmer Dougal Williams.

Douglas Cook's father William Cook, born 1855 at Herry Hill, Aberdeen, Scotland. (Photo Jan Potter)

Douglas Cook's mother, Jessie, (nee Miller) born in 1853 at Sharland's Villa, Sharlands, Glasgow, Scotland. (Photo Jan Potter)

of developing Pukekura Park which was to become one of New Zealand's notable gardens. Douglas did not grow up in New Plymouth but he developed a love of horticulture as a schoolboy and he must have revisited his birthplace from time to time. His memorabilia in later years included early photographs of Pukekura Park and notes he had made on its development. And he was always enthralled by the symmetrical beauty of Mt. Egmont (now Mt. Taranaki) and the luxuriance of its tree ferns. Jack Goodwin, former director of parks in New Plymouth, wrote in 1967 that Douglas Cook's photographs and notes on Pukekura Park showed "a glimpse of the enthusiasm that was to make him one of New Zealand's most knowledgeable plantsmen".

To the prodigious task of converting 620 acres at Ngatapa into productive farmland, and ultimately into a place of great beauty, Douglas brought a Victorian upbringing and a family motto "On the family crest is the word Forward and I determined never to look back". That motto was to remain a driving force through his most difficult times and sustained him until his death.

He was the second son of a bank officer, William Cook, and Janet Turnbull (Jessie) Cook, nee Miller. There was a family background of affluence, of Presbyterian religious values, of refinement and of adherence to Victorian codes of behaviour. Though, as a teenager, Douglas was to rebel against the domination of his father, he certainly retained a taste for the refinements associated with a "country gentleman". He also inherited appreciation of books, learning and the arts. (His mother had been educated in London, Paris and Bremen.)

His paternal grandfather, John Cook, had been a shipowner in Aberdeen; his maternal grandfather, William Miller, was a city councillor of Glasgow and was a partner with Sir Archibald Orr-Ewen in cotton mills in Kilmarnock and dye works in Glasgow. It would seem that William Cook and Jessie Miller were betrothed before William left for New Zealand aboard the Farnsworth, arriving at Auckland on September 8, 1879. He took up a position with the Bank of New Zealand the next day. Jessie followed almost two years later, landing at Auckland from the vessel Hermione on August 19, 1881. The couple wed three weeks later. Their first son, John, was born in Auckland on September 20, 1882. During a period of frequent transfers to different branches of the bank in both the North and South Islands, the Cooks moved to New Plymouth where Douglas was born on October 28, 1884. A daughter, Sheila, was born in Auckland on Christmas Day, 1891.

Douglas was to have his first journey abroad as an infant when his mother took her two sons to Britain aboard the Aorangi in 1888. The young Douglas Cook must have had his schooling interrupted frequently by his father's transfers. However, he was in his fourth form year at North School, Oamaru in mid-1898 when he left there to enrol at Wellington College. He was aged 17 when he left school in November 1901.

Douglas loved and respected his mother, but his relationship with his father was far from warm. He wrote in 1964: "My mother was the sweetest, gentlest little lady, always tidy and well-dressed but not extravagantly. She was brought up with everything a wealthy father could give, but my father was just a bank accountant and an awful spendthrift. He invested in the maddest things and lost all. My poor mother was very unhappy but showed little. My father liked my brother, but not my sister or me. He held the floor; a very conceited and self-opinionated man." And later, after a visit to England: "I wish my father had been alive to hear of the heights to which I rose. I, the son he used to call awful names. Filthy little skunk, brainless creature and lots more. I didn't have the guts to say, 'I've inherited them all from you'."

When Jessie Cook died in Auckland in 1935 in her eighty-third year, a newspaper obituary described her as "deeply religious and of a shy and retiring nature". Her husband predeceased her by seven years.

Tension between father and son appears to have been the reason for Douglas leaving home at the age of 17 – in his words, "I cleared out from home and school and found myself a job". He began his working life as a cowboy in Hawke's Bay for Hughie McDonald who leased Stirrett-Orr's Rosemount Station at Puketapu. The following year he must have been on improved terms with his father when he borrowed from him in order to buy a block of the Frimley peach orchard on the outskirts of Hastings. The Frimley homestead, large orchard and an associated cannery had been established there by James Nelson Williams, a noted pioneer and land developer – coincidentally, the grandfather of H.B. (Bill) Williams who, 60 years later, was to become a crucial figure in the fulfilment of the Douglas Cook dream.

Of his Frimley enterprise, Douglas wrote: "At 18 I borrowed £140 from my father and bought 10 acres of Frimley Orchard and went shares in a three-roomed cottage... My friend was Charles Douglas. We had a bedroom each and a sitting-room, kitchen in between and a lean-to behind to keep the rain off our one door. We had no bathroom but bathed in a big tin tub behind the house but in full view of Omahu Road till I, being shy then, erected a 6ft scrim on four pegs round the bath and artesian pump. In winter we brought the bath on to the rug in front of our stove and poured two kerosene tins of hot water into it and bathed in the sitting-room.

"Charlie was a wonderful, fine character and he taught me, by example, good manners and, though I didn't realise it,

Douglas Cook on a Sunday picnic at Takapau Station, near Waipiro Bay, about 1908.

how to speak decently. He later was secretary of the Hawke's Bay Show and then secretary of Hawke's Bay Farmers, finally manager of Hawke's Bay Farmers. He was a Presbyterian, as I am, and a very upright, honourable and good man. He, more than my parents, brought me up and taught me manners and how to be a good man. I have ever thanked him for the benefit he was to me. I have tried to live the honourable life he taught me. It's been difficult; often sometimes I've slipped. None of us is perfect."

The two friends experienced adversity together as frosts wiped out their peach crops in two out of four seasons. They left their peach trees to find work on farms and Douglas moved up the coast, north of Gisborne, to Waipiro Bay where J. N. Williams had been breaking in big acreages of farmland. There is an early photograph of Douglas on Takapau Station, inland from Waipiro Bay, and a story of him upsetting A. B. Williams, one of J. N.'s sons. Douglas retreated into the scrub, returning after two days to gather his belongings and leave. He returned to Hawke's Bay and on being successful in the ballot for a section of the Ngatapa Settlement he sold his peach block for £583/10/8.

To his farm he brought enthusiasm, energy and pride: "I never could stand taking orders and loved roaming the hills. When I came here first at 25 I was happy as a king and as strong as a bullock. Work of any kind was no trouble. I used to sing on the hilltops at the top of my voice, often home-made words and often, too, words of praise and thanks to God for my happiness of being free and the owner of lands. Mine to think out the future of! Mine to plant as I wished!"

He named the property Eastwoodhill after the Miller family home at Thornliebank, near Glasgow, thus creating a truly living memorial to his mother's life. In later years he made a pilgrimage to his mother's old home: "It is now an old people's home and I loved seeing it almost as my grandfather left it. I gave instructions to a nurseryman to arrange to make a deciduous azalea bed and plant a dozen well-budded plants before spring. That would help brighten the old people. My grandfather would have been pleased his lovely place was a comfort to old people."

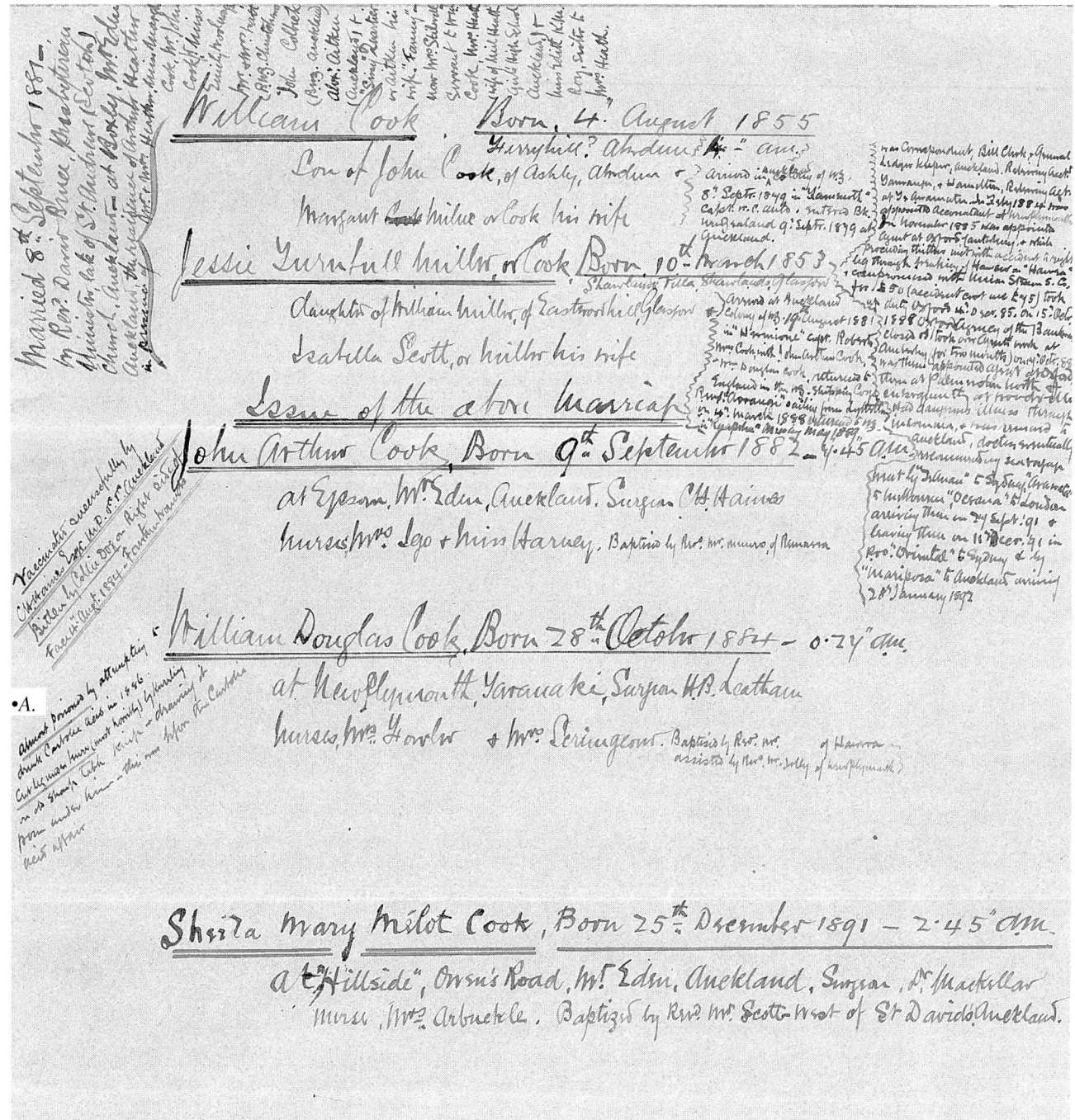

A handwritten note from William Cook records details
about his marriage to Jessie, the birth of their three
children and some work and travel details.

•A. Details relating to Douglas Cook in his first years.
"Almost poisoned by attempting to drink carbolic acid
in 1886. Cut leg under knee (most horribly) by kneeling
on old sharp table knife - drawing it from under him -
this before the carbolic acid affair"

2 Birth of a Park

There were no buildings of any kind on the Douglas Cook property when he arrived in 1910. His first decision was to choose a hilltop for his house site and, while staying with the nearby Monckton family on their Hylands Station, he had a simple cottage built. Initially he had only three paddocks on the farm, one of them being over 600 acres and he had to set to work fencing, and clearing the scrub. No one was ever to describe Douglas as a "practical farmer" but in his first plantings he did follow a utilitarian approach – some mixed eucalyptus and *Pinus radiata* for future firewood, and an acre of mixed orchard including apple, pear and plum trees for household use. Predictably, however, decorative planting quickly followed: "I had always loved beauty around me and in less than six months had quite a garden surrounding my one-roomed shack." Friends gave him "bits and pieces", almost all of them suckering plants.

Douglas Cook set to work subdividing the 600 acre paddock of the old Ngatapa Station. The photo is taken by the Taumatapoupou Stream not far from the present Hihiroroa Road bridge where the eastern boundary fence has been erected. Pictured is J. Banham who started work at Eastwoodhill in March 1911.

A panorama from Orchard Hill in May 1910 six weeks after Douglas Cook took up residence. This area now incorporates Pear Park and the Homestead Garden. He has already planted a garden about the cottage including the daffodils he bought when he was at school. A manuka brush fence gives shelter from north-west and southerly winds.

The "Big Work" was the hurried planting of the trees, shrubs, rhododendrons and azaleas in the area to the south of the House Terrace. Andy Cole (pictured) helped right through this time. This was in the last days before Douglas Cook travelled to Dannevirke to enlist for service in World War 1. At the time his father was managing the BNZ at nearby Pahiatua.

In the winter of that year he set a pattern that was to continue throughout his life by ordering 75 rose plants from an Ashburton nursery. His aim was to develop land around his cottage as shrubberies, plantations and park lands and "to plant as much of the valley below the house as I owned in such a way as to beautify my view, without losing much grazing for long. The horizon down the valley was 60 miles away and all I had to deal with was the first mile".

The outbreak of World War 1 in 1914 abruptly halted the development of Eastwoodhill. Douglas volunteered to serve. Only days before he was to enter training camp, an order of 100 trees and shrubs and 100 rhododendrons and azaleas arrived from a nursery. He had to plant frantically before his departure. Among those trees were three oaks. He later described them, writing in the New Zealand Gardener magazine in May, 1950: "Two are crimson red and the pin oak is rapidly turning to its usual browns, yellows, golds and some red. For three weeks the red oaks have been good and, given peaceful weather, will hold beauty for another fortnight. These were planted in 1914, a week before I left for camp, and are lovely trees now."

He enlisted in the Wellington Mounted Rifles at Dannevirke on August 17, 1914 and embarked for the Middle East, with the rank of trooper, on December 3, 1914. In action at Anzac Cove, Gallipoli the following July he lost the end of a finger to an enemy bullet. There was service in Egypt before he was transferred to the 10th Battery, N.Z. Field Artillery as a gunner in March, 1916. The war for Douglas ended in France: Only five days after being promoted in the field to the rank of bombardier he was wounded on September 26 in the face, neck and left forearm. He was left partially blind and spent time convalescing in England before he was invalided home, unfit for further service, departing aboard the S.S. Remuera in March, 1917.

Making your own was an essential part of life at Eastwoodhill. J. Banham shows his cabinet-making skills, in about 1912, at the workshop where the Douglas Cook Centre now stands. The workshop was later moved adjacent to the present pavilion site.

He later wrote: "Came back blind in right eye. Couldn't draft sheep; couldn't see their earmarks. Couldn't drive a staple or nail but worst, most hampering of all, couldn't drive a car. I didn't realise at the time how I hated war but, when I got home permanently unfit for further war service, soft and unfit, I used to sob at night, thinking I'd never be a strong man again. Work and plain food soon brought back my old strength. I was shocked on my return at the social life and spending of money when the soldiers were living and fighting, and often more."

He enjoyed, over the ensuing years, the company of fellow veterans, particularly at reunions, but avoided talking to anyone of his wartime experiences. There was one remarkable exception. In 1936 his 14-year-old nephew Peter Wily, son of Douglas's sister Sheila, was visiting Eastwoodhill from his parents' farm near Pukekohe, south of Auckland. Peter had a good relationship with "Uncle Douglas". The lad respected the property and was careful not to disturb any plants. Sometimes he would join his uncle and assistant in the woolshed at "smoko" time. Best of all, he was "treated like an adult" in the dining-room. He sat proudly at the big table even for formal dinners with as many as eight or nine adults, though the discussions were over his head.

"Douglas seemed to regard me as a friend despite the difference in our ages," Peter recalled. "We talked quite a bit." One day his uncle told him a poignant story about an encounter in France with a dying German soldier – an event that had deeply saddened him. It was a story that Douglas had never told before.

In 1918 a horse-drawn wagon is delivering three tons of trees and plants for the Eastwoodhill garden including a bundle of Eucalyptus. Most tree purchases were at this time coming from Horton's Nursery in Hastings. The bare skyline hill at the upper right is above the cabin in Cabin Park, part of a sandstone seam that crosses Eastwoodhill. From the left peak, now called Bishop's View, a superb view can be had over the tree tops of the Circus, 200 feet below. The scene pictured is on the present entrance drive just over the culvert where today cars drive on tar-seal and surrounding trees obscure distant views.

Douglas Cook's war service was to have, in an indirect way, a major impact on the development of Eastwoodhill through his convalescence in Britain: "I stayed in several beautiful country homes in Britain and this left me with a growing desire to create something worthwhile in New Zealand… I'd got the idea that I too could have lovely surroundings, even if I could never have a fine home and live as they did. That was the start of the park. A dignified park to drive through to my home, whatever its size." He referred to a family saying, "Set your heart on an objective and believe in your heart you'll achieve it and you will. Where most men fail is lack of faith in themselves." His vision for Eastwoodhill had expanded and he acted almost immediately. In his own words, the planting of Eastwoodhill "started in earnest" in 1918.

During his period of convalescence he had visited the Royal Botanic Gardens, Kew and, perhaps through the assistance of his influential relatives, was able to meet Arthur William Hill, later Sir Arthur William and Kew director who gave him a plant each of the red and variegated cabbage trees. Douglas carried them home in his pack, thus introducing those varieties to New Zealand. More than 30 years later, Douglas admired the way the purple cabbage tree "showed up" against a weeping beech.

Many thousands of *Pinus radiata* were planted on dry ridges and other "pumicey, poverty-stricken areas" for timber and future firewood. *Eucalyptus viminalis* was planted in blocks 6ft by 6ft for rails and *E. macarthuri* for posts. "When utility planting was finished we turned our attention to the garden and made a start on park planting. *Platanus orientalis*, various elms, *Acer pseudoplatanus*, *A. platanoides* and others were used 60ft apart along road fences and a few as specimens in the paddock... On our old drive in 1918 I planted an avenue of Lombardy poplar and to get the effect I wanted in my lifetime, I planted the trees 6ft apart in rows and each row 12ft from the centre of the drive." At the time of his writing, in 1949, he added: "The trees are now quite 70ft high, maybe 80, and the golden glow in autumn is wonderful, not to speak of the cathedral effect."

The soil in which Douglas Cook was creating his park posed problems. Above the sedimentary sandstone and mudstone, volcanic eruptions in the central North Island during more than 50,000 years had showered the area with fine ash and pumice. The weathered sandstone is able to hold moisture, and once a tree is established, with its roots reaching down to that layer, it can survive. The ash layers, however, dry quickly in Gisborne's dry summers. The only source of water on Eastwoodhill available to Douglas Cook in the early years was the small and seasonal Taumatapoupou Stream that ran through the property. There was a lack of dams or any other water supply once the stream was dry. Douglas used mulching techniques. He also adopted a policy of watering trees for two years, after which they had to fend for themselves. Until ponds were established in the late 1950s and early 1960s there were

Douglas Cook's first garden plantings between 1918 to 1926 were in a long narrow strip down to the Cabbage Tree Avenue. This view is within the present Homestead Garden between the tractor sheds and the lower lawn. Some of the plantings still exist, notably on the right a deodar cedar and Norway maple. Although there are shrubs as well as trees, sheep are already being used to keep the grass down.

The poplar rust decimated the Lombardy poplars of the Poplar Avenue in the early 1970s. Originally the main entrance, it was replaced in 1936 by the 'new drive'. It is nowadays again the main entrance for traffic safety reasons. (Photo Douglas Elliott)

heavy losses of young plants – more than 50 per cent failed. It was a measure of his dedication that he would not allow such a difficulty to stunt his dream.

He was to describe the climate of Eastwoodhill as about as genial as that of the mildest parts of Britain but with much more sunshine (on average 2200 hours) and much less humidity. Unlike Gisborne on the coast, Ngatapa experienced some sharp winter frosts as Douglas was developing his park – "As the garden starts at about 400ft above sea level there is quite a range of climate. If a tree, shrub or plant is frost tender in one area we move it higher up the hill till it finds a spot in which it will thrive. In a garden of hills and gullies it is amazing the problems there are. It is a lot of fun succeeding eventually with a thing that has beaten you for years." No ailing plant with a possibility of survival was wasted: "What appear to be duds are not thrown away here; they are moved to less conspicuous positions and given a chance."

Substantial orders were made over the next two years – 1996 trees and shrubs in 1919 and 3387 trees and shrubs in 1920 of which 2500 were for the garden. The rest were plantation trees. He wrote: "In 1920 a very large consignment of trees and shrubs was brought in from Hawke's Bay, but the weather broke when half had been delivered and the balance had to be sledged and brought in on pack horses for the last five miles over roads churned into a sea of mud. There were no tar-sealed roads in New Zealand then, unless in Taranaki, and the country roads in our district were clay or just earth.

"Wonderful days followed; 8ft high specimens soon made certain areas look park-like and what we then thought a large and varied shrubbery came into being. Looking back, how poor those shrubs were, how small the area. Yet we thought them wonderful – they were the best available in New Zealand in 1920."

He engaged a Mr Hughes to plant trees, mostly plane and elms, along the Wharekopae and Hihiroroa roadsides skirting the property. They also planted apple trees along the Wharekopae Road to provide fruit for passing drovers. The local roads were not metalled until about 1927. The winter mud made it impassable to horse-drawn vehicles so bullock teams were used to haul freight to the settlements of Rere and Wharekopae. Douglas allowed them to camp overnight in a roadside paddock. With the noise of the animals and that of men singing and drinking in the evenings, Douglas described the scene as being a "real circus". He subsequently named that roadside paddock, The Circus.

In those days before tree roots had ramified the soil, bulbs and perennials found friendly ground beside paths where the soil had been dug out and laid to a greater depth. The orders of 1920 included paeonies, tulips, hyacinths, lilies and gladioli.

An inventory of the 1920 plantings with material from a variety of nurseries showed a total of £366 spent.

MONEY SPENT 1920 ORDERS

Trees & Shrubs

Hortons (Hastings)		£99.08.00
Duncan & Davies (New Plymouth)		£22.14.09
Ivories (Rangiora)		£20.00.00
H.B.Just (Palmerston North)		£19.12.11
Hayn (Auckland)		£ 0.07.10
Matheson & Roberts (Dunedin)		£3.00.00
Ant Roozen (Holland)		£10.00.00
A. Montague (Camberwell Victoria.)		£4.05.00
H.C. Gibbons (Wellington)		£2.00.00
Gilbert J. MacKay (Auckland)		£6.08.09
Anderson (Napier)		£5.00.00
Cooper (Wellington)		£0.07.06
Bees (England)		£2.00.00
	Sub-total	**£195.04.04**
plus		
Freight		£60.00.00
Cartage		£28.00.00
Telegrams, telephoning & postage		£2.15.08
plus		
Japanese order all costs		£60.00.00
Ant Roozen Paeonies, Tulips, Hyacinths, Lilies & Gladioli		£20.00.00
	Grand-total	**£366.00.00**

It was a staggering sum in terms of the sterling value of the time, and compared with normal farming practice. No wonder Douglas Cook came to be regarded among some of the local farmers as rather eccentric – a notion fuelled by the fact that he enjoyed working in his garden without the encumbrance of clothes. Douglas, on the other hand, never regarded himself as a conventional farmer. It surprised him that the "average farmer" was satisfied with his wife's half acre of garden and seldom planted anything other than pines, willows, gums, wattles and an occasional poplar himself. He reflected: "What a pity more of them do not take a little more interest in the type of tree they plant so that utility might be combined

At the corner of the Wharekopae and Hihiroroa roads by the present exit gate, Mrs Sutton, housekeeper for Douglas Cook 1919 and 1920, reads a letter from the mail bag. On the left, over the fence, is the Dry Ridge which lies between the present entrance and exit gates in Corner Park.

with beauty." To Douglas, there were decorative trees which could perform practical functions; for example, those decorative varieties which could replace willows for stream sides and light up the country with their glory. "Perhaps some day we will be sorry we did not do more to beautify our country when times were good."

No planting was done in 1922 or 1924 when Douglas was travelling in Europe, but during both those years he saw "really good things" at Kew, at the Chelsea Flower Show and in the gardens of the country homes where he stayed, some of them 300 years old: "I loved the country scene in England and its beautiful parks, and wondered how far I could get in the creation of a park in one lifetime. I determined to make a start."

In 1926 the garden was greatly extended. He earlier imported thousands of tulips, hyacinths and paeonies from Holland and created broad herbaceous borders.

There is firm evidence that Douglas Cook's initial intent to create on his property something of the stately beauty he had seen in Britain had expanded further during the early 1920s. That evidence came from Lawrence Goodman, whom Douglas had met on board a coastal ship en route from Auckland to Gisborne in 1923. They became friends. Lawrence had disembarked at Tokomaru Bay, on the coast north of Gisborne, but on occasions he travelled to Eastwoodhill to stay with Douglas. Sometimes they worked together in the garden. Lawrence Goodman recalled, years later, that in 1923 Douglas told him that his vision was "to build a garden for all New Zealanders". Douglas was a prolific letter-writer and that vision was to become a recurring theme throughout his life. In 1948 he wrote in the New Zealand Gardener of the need for someone young to take over his job and get to understand his aim – "Someone with energy and some money, with all his life ahead of him and who does not make money his aim. Someone who can give, as I have given, to create something beautiful, some place restful in a world full of hates and greed…"

3 Right-Hand Man

When a young farm cadet from Flock House in the Manawatu arrived at Ngatapa in 1927 he walked straight into the pages of Eastwoodhill's history. At the time, Bill Crooks could not have known that the park would become both his work and his home for virtually the rest of his life. Faithful and loyal, he was to serve Eastwoodhill for 47 years. He was much more than a farmhand; he served over the years as assistant plantsman, chauffeur, drinking companion and, above all, right-hand man.

Flock House had been established through the efforts of the Member of Parliament for Rangitikei, the Hon. Edward Newman, who wanted the farmers of New Zealand to acknowledge the efforts of the British seamen who kept the sea lanes open during World War 1. In conjunction with the Sheepfarmers' Association of New Zealand, he established a fund so that the sons of British seamen who had been killed or wounded could be trained in New Zealand to start a new life on the land. A property was purchased in 1924 and Flock House

Some of the rhododendrons planted in 1914 required more room by 1928. This one is being moved down to near the front gate in Corner Park alongside the ten-year-old Poplar Avenue which is the present entrance drive. On the left is the young Bill Crooks.

A 1930 photo of Bill Crooks showing garden development. From the sundial to the bulb beds, Douglas Cook erected an arch planted with wisteria. In it he installed a cold water shower for relief on hot days.

remained an agricultural training facility, with a broadened base of admission, until 1988.

Bill Crooks was a slightly built youngster, only 5ft 2 ½in tall and weighing 8st 71b when, as a 16-year-old, he entered Flock House with the first intake. The son of James and Jane Crooks, he was one of seven children born in Lowestoft, a fishing port and resort town in Suffolk on the extreme eastern coast of England. Lowestoft is famous for its herrings, and James Crooks was a "smacksman" (skipper of a fishing smack). During World War 1 he was appointed to the Trawling Reserve of Britain's seagoing defences. He won the Distinguished Service Medal for sinking a German submarine by ramming it with his trawler and his name was honoured in Lowestoft "as one of the bravest fishermen of the fleet". James Crooks died on December 27, 1919 from pneumonia, aggravated by wartime injuries.

Bill had left Roman House School early in order to work for a wine merchant and to supplement the family income. Then, with the opportunity of entering Flock House, he emigrated with his 15-year-old sister Gertrude. They sailed on May 24, 1924 aboard the SS Remuera – the same vessel that had brought Douglas Cook home from the war. "Gertie", as the family called her, entered domestic service. Flock House records of the time described Bill as "honest and cheerful and extremely keen to make a way for himself from farming". After a year's training there he worked on the Sherratts' Horoeka Station, west of Gisborne, and is believed to have spent some time at Whakapunake Station near the inland settlement of Tiniroto, before entering Douglas Cook's employment early in 1927.

Farm workers came and went at Eastwoodhill. Douglas Cook was not a patient man; indeed he was known to neighbours for sudden outbursts of ill temper. It was a measure of Bill Crooks' loyalty that he survived all such storms, though it is clear that he often had cause for complaint. Bill married Josephine Richardson who was the daughter of a pioneer farmer and had grown up at Wharekopae. She and Bill raised a family of four sons and a daughter while living in a small, two-bedroomed cottage on the section of Eastwoodhill known as Pear Park. Their lives were far from comfortable. They never had better than an outside toilet and their laundry facilities

consisted of a copper, outside under a lean-to. Their water supply was poor and sometimes, when it ran out, Bill had to siphon water from the Lawn Cabin tank, hoping that Douglas would not notice.

Douglas, with his blind right eye, was totally dependent on Bill as "chauffeur", and when Bill and Jo were planning to buy a car for themselves he shrewdly went half shares with the couple. It meant that whenever Bill and Jo were going out together Douglas would be there too – as the front-seat passenger with Jo in the rear. Eventually Jo gained a degree of independence when she and a sister bought a car between them. But there was no all-weather driveway to the cottage and Douglas would not allow her to drive her car up the main entrance to Eastwoodhill. Thus her parking the car in Windy Gap by the Daffodil Patch in wet weather for loading and unloading was "illegal" according to Douglas's ruling. She was expected to use a more difficult slippery route through Pear Park whatever the state of the weather. Douglas's quirk of working in his garden clad only in hat and boots was also a problem for Jo. She would not venture far from the cottage without knowing where he was working. The children were not allowed to keep a horse because it would eat Douglas Cook's grass.

Looking north across the Wharekopae Road in late 1931 to the newly-planted Corner Park, with the Poplar Avenue on the left. The photograph was taken after the house fire with the first section of the new homestead showing. Douglas Cook's woodlots of radiata pine, Douglas-fir and eucalyptus feature prominently. Towards the right, there is a shepherd's whare. The dark area (centre left) is the Daffodil Patch with Wet and Elm Lawns (centre right).

*Bill Crooks beside the Hillman Minx bought in 1938
for £318/13/6. When Bill and Jo Crooks decided to take
a town day, Douglas Cook would be there and in the
car. After a while he told Jo that she need not go to
town every week, once a fortnight would be enough.
This was later extended to once a month. As a result,
Jo and her sister Phil Hindman went half shares in a
car which gave Jo independence from Douglas's whims.
However, he placed restrictions on which drive she could
use to get to the cottage.*

Despite such irritations, there was also warmth in the
relationship. The couple named their first-born after Douglas
–William Douglas Crooks – and Douglas was the boy's
godfather. Douglas Cook took an interest in the children over
many years. He was a stamp collector and introduced young
Douglas to the hobby, showing him how to soak stamps off
envelopes and how to dry them. He gave him packets of stamps
and when the lad attended Gisborne High School in town as a
boarder, he sent him postcards from exotic locations during a
trip on the Continent in 1954, carefully noting items of
geographical and historical interest. One from Syracuse: "I'm
standing outside the catacombs while the mob go in to see
hundreds of dead bodies, standing, sitting, lying and looking
at you. I saw them at Palermo and that was enough for me.
You can find this place on the East of Sicily, south of the Straits
of Messina. We go from here to Cadiz, near Gibraltar in Spain."
He signed himself W.D.C.

Another reflected his taste for grand living. Featuring a
photographic print of the handsome Villa Taranto at Pallanza,
Italy, with lush gardens and a fountain, the card said: "I have
been here in 100 acres of garden for four days. The Villa is like
a palace. Three men wait on table and I have a valet to myself.
A very delightful way of life for a few days. The owner is a
Scotchman (sic) and a millionaire. Has given the property to
the Italian Government but lives here for life… Best wishes.
W.D. Cook." (His host was Neil McEacharn.)

White peppermint gum, Eucalyptus pulchella, *at the main entrance.*

Stephen Jones

*William Douglas Cook, an unfinished portrait by John
Wheatley, Associate of the Royal Academy, 1955.*

ii

Heathcote Beetham (Bill) Williams, 1996.

The St Landry oak, Quercus x ludoviciana *on Crooks' Walk, Pear Park. Received in 1951 from Hiller's Nursery, it was 1.25m in diameter, 24m in height and had a spread of 36m in 1996.*

The arrival of Bill Crooks in 1927 had an immediate impact on planting and his presence made possible the creation of a true tree garden. While Bill attended to farming chores, Douglas Cook had more time to devote to planning, writing out orders to nurseries and punching out the lead labels he used to identify new plants. Bill also assisted with planting.

"It was not until 1927," Douglas later wrote, "that we really started into park planting. So much wrong planting had already been done that we decided to take in the whole area between the house and the road and, year by year, plant what we had time for and work to a plan of the area." That section was to become known as Corner Park. They laid out paths, and planted a two-acre slope that was to become the Daffodil Patch. Oaks and poplars were planted along the Hihiroroa Road in August, 1927.

The late 1920s were also notable for the arrival of a new friend for Douglas Cook. There was something of a storybook quality about his meeting with Ethel Tidswell, a young schoolteacher who was to become hostess for his dinner parties, as well as a good companion. She was riding a horse on Hihiroroa Road; Douglas was working with Bill Crooks down by the woolshed. He raised a hand in greeting.

"You'd be the new schoolteacher," he called. Ethel rode up to the fenceline and they chatted. Ethel was in her twenties, Douglas was 20 years older. Ethel knew of him because he was a friend of the White family with whom she was boarding. What interested her was his knowledge of and involvement with plants.

"During my studies, even as a girl, I had always been fascinated by botany," she was to recall. "We clicked."

Ethel was born in a large family home opposite the Makaraka racecourse on July 30, 1902. Her father, Joseph Tidswell, was a prominent woolbuyer, selling wool for Douglas Cook on occasions – a Yorkshireman and "former Bradford gentleman" who used to write newsletters for The Times of London about the prices at New Zealand wool sales. He was also a man with a respect for trees. At centre front of the Makaraka property he planted a Norfolk pine. It flourished and became a local landmark. Many decades later, when Gisborne aerodrome was being extended, the pine was so tall that it was considered a danger to aircraft, and was felled.

One of several thousand lead tags punched out by Douglas Cook. The thumping of the hammer could be heard echoing long into the night. He punched labels according to the orders he had made, so if the plants did not arrive those labels were then spare. Apart from those still on the trees, there are over three and a half thousand in the label 'cemetery' today. The code at top left is the supplying nursery, in this case Hilliers of Winchester in England. The figure at upper right gives the eventual height of the tree in feet. The botanical name is in the centre, and the country or geographic region to which the tree is native on the lower edge. They were fixed to the tree with a copper wire looped about a branch. Their actual size is 1¹/₂ by 4 inches.

The Taumata School was sited alongside the Wharekopae Road on a portion of Morley Stafford's land about 2km on the Gisborne side of Eastwoodhill. A shelter belt of pines and a single oak remained until the November 1994 winds. In this photo, prior to Miss Tidswell's time, are two of Dougal Williams daughters and three boys of the O'Connell family. The main portion of the school was later shifted to Jock White's backyard in Gisborne. (Photo Morgan O'Connell)

Ethel was "brought up as a lady" and was not expected to go to work. At the age of 17 she went to the local de Lautour sisters as a form of finishing school – "I was taught manners and how to pour cups of tea." However, she decided to become a schoolteacher and gained a challenging position "closing up schools". Where schools had become redundant through declining rolls, entry would be closed and Ethel would teach the remaining older pupils through to proficiency standard. At rural schools she boarded with the parents of pupils. At Taumata School, about 2km east of Eastwoodhill, one of the senior boys was Jock White and Ethel stayed with his parents on week-nights.

When Ethel met Douglas Cook and began to have regular meetings with him a few tongues wagged. His quirk for working in his garden in the nude was well known in the neighbourhood. One matronly countrywoman warned her, "You mustn't go there!" Ethel, however, enjoyed his company and particularly relished the opportunity to learn more about plants. Douglas (fully dressed) taught her how to lay out the roots when planting a seedling and how to stamp it in. He told her how to water them – "Most people water too much" and how to choose plants that would be compatible with the soil. She borrowed many of his botanical books.

"I used to help with the digging," Ethel recalled. "I was a big, strong woman."

The issue of nudism never became a problem. Sometimes Douglas would tell Ethel, "I'll be naked tomorrow" as a warning against an unannounced visit. Ethel knew that some neighbourhood boys used to try and spy on him "but Bill Crooks was good at chasing them off". On Fridays after school she used to drive the 35 km home to Gisborne in her gig, drawn by her show-class pony, Tilly, to spend the weekend with her parents. She told them of her friendship with Douglas; they did not protest – "I was proud that they could trust me in such a position". At first, Ethel always addressed Douglas as "Mr Cook" – "We were never too familiar; we kept our distance." Later, as the friendship developed, she sometimes called him "Cookie" – "but I always called him Mr Cook if I wanted to impress anything on him ! " It would have been a bonus for him in their friendship that Ethel could cook, bringing him freshly-baked scones or buns on her visits. Once he told her:

"It's time you came and cooked me a decent dinner…" She obliged with a roast, with all the trimmings.

She was to assume a more important role: Douglas had quite a circle of friends in the farming community and enjoyed staging dinner parties, employing a full range of his crystal glasses and silverware. Ethel became the hostess. She would stand beside Douglas as he welcomed the guests at the door, but he did not formally introduce her. There were some extraordinary rules. One was that she was to wear a plain evening dress, not as grand and flouncy as the lady guests could be expected to wear. Before the visitors arrived she was to push all valuable ornaments to the back of the mantlepiece and other shelves. And she was to count all the glasses. It seems most unlikely that Douglas would have expected any of his guests to steal one; perhaps it was his way of keeping a tally on breakages. Replacements were ordered from Stuarts in England. While the glasses had the characteristic Stuart stem featuring four hollow spiralling strands, Douglas selected the Waterford pattern for the bowl.

On Ethel's first night as hostess there were breakages. The host did not stint on supplies of wine and other liquor. One lady guest slumped in her chair, oblivious to the proceedings, before the night was over. Ethel did not drink alcohol but poured drinks for the guests. She slept in the house that night because the party went on into the small hours of the morning. No doubt she was weary after washing all the glasses, counting those that were still intact and putting them carefully away.

There were many dimensions to Douglas Cook. Ethel was deeply impressed by his love of books and his wide range of study-reading: "He had a wonderful brain." She recalled that before going abroad he would spend numerous hours poring over books about the destinations.

She also admired his worldliness and his sophistication; the fact that he had a hand-stitched suit made by a tailor in London.

The Taumata School closed and Ethel moved on. At the age of 93, in the year 1995, she looked back on the Ngatapa years and observed: "Throughout our friendship he was a gentleman – and that covers everything."

Mrs Kate White, with whom Ethel had boarded, was

A water jug and some glasses of Douglas Cook's Stuart crystal.

Douglas Cook had a range of fancy dress costumes which he paraded proudly on occasions when he hosted parties. There were Christmas parties in the woolshed or the Daffodil Patch where Guy Fawkes celebrations were also held. Jo Crooks told of a Christmas Eve party in the Daffodil Patch where the men had decorated the **Picea omorika.** *At the end of the evening the Crooks family were told to take down all the decorations before they went home, something of a damper on the thrill of the evening.*

among Douglas Cook's closest friends. Mrs White, in fact, was one of the few women who established a true companionship with him. Douglas would shower all lady guests with charm and courtesy but, in the tradition of Victorian gentry, tended to reserve serious discussion for men. Mrs White's daughter, June Murphy, was to comment in later years: "Douglas was fond of mother. She was a practical, commonsense person who never prattled or gossiped, and Douglas Cook never talked about people." Mrs White was also one of the few who, on informal occasions only, would call Douglas "Cookie" rather than "Mr Cook". He once quipped to a friend, "If you wish, call me Cookie but never call me Doug. I can't stand a good name like Douglas murdered."

The White children – Jock, June, Richard, Ted and Ian – were among the neighbourhood children who attended Christmas and Guy Fawkes parties elaborately staged by Douglas. One Christmas he dressed in a Mandarin costume to greet his young guests. Each child was given a numbered ticket corresponding with a gift on the Christmas tree, a decorated conifer. Guy Fawkes celebrations were noted for the lavish fireworks displays, including spectacular Catherine wheels. Douglas's generosity extended beyond the parties; he enjoyed giving gifts. When one young woman became engaged he invited the couple to choose their own wedding present from three items he had arranged on the dining-room table. He left them alone while they made their choice.

When Douglas felt that there was something of particular beauty or interest in the garden he would invite neighbourhood farm wives, among them Mrs White, Mrs Grace Monckton and Mrs Heni Sherratt, to afternoon tea. They, of course, would bake cakes and other goodies for the occasion. Douglas was nervous about having young children in the house, but as June grew into a teenager she was privileged to take a place at the table, splendid with its fine china and George I silverware. There was always a choice of India or China tea, made from tiny buds which Douglas kept in special containers. One day Douglas invited June to pour the tea – "It was a great honour and a great pleasure. It was his way of honouring the fact that I had grown up."

4 A Wife and Son

Douglas Cook was visiting his sister, Sheila Wily, at her Pukekohe home in 1929 when he met another guest, Claire Bourne. She was assistant librarian at the Auckland University Library and, with Douglas's passion for books, it is not surprising they struck up an acquaintanceship that was to blossom.

Claire, who was born in Christchurch, was the younger daughter of Charles Frederick Bourne, formerly of London, who had been headmaster of Auckland Grammar and later of Christ's College, Christchurch. Her mother was formerly Margaret Roe and, through that family, Claire was a cousin of two British pioneer airmen, the brothers Alliott Verdon Roe (later Sir Alliott) and Humphrey Verdon Roe, founders of the Avro Aircraft Company. Claire had been a teacher for several years before becoming a librarian.

After a four-month engagement Claire and Douglas wed in St. Aidan's Church, Auckland on October 20, 1930. She was aged 36; he was her senior by 10 years. Though life at Eastwoodhill must have been strange to Claire after her professional life in busy Auckland, she quickly fitted into the scene. Neighbours welcomed her and enjoyed her company. She liked the open spaces, the sounds and vistas of the Ngatapa countryside. She enjoyed carrying the "smokos" to whichever section of the property Douglas and Bill Crooks, and sometimes other helpers, were working. She had not been interested in gardening before her marriage but she proudly established, and nurtured, a vegetable garden beside the cow bail. Douglas, however, had no time for vegetable gardens, feeling that they did not fit in visually with the trees and flowers. Uncharitably (but probably not within her earshot) he spoke of her "scratching like a hen" efforts in her garden.

Disaster struck in the first year of their marriage. Douglas had been known for absent-mindedly leaving food on the stove while he became absorbed in the garden. On a day that Claire was absent, having travelled to town by the Rere freight car to see the dentist, Douglas and Bill were digging potatoes down by the front gate. They smelt smoke and raced up to the house. It was in flames and beyond saving but they were able to throw an important chest of drawers out of a bedroom window. It contained 10 years of planting records as well as photographs of the park dating from 1910. But Douglas lost books, works of

Douglas and Claire Cook at Eastwoodhill soon after their marriage in 1930. (Photo Sholto Douglas Cook)

Douglas Cook's wife Claire with their cat (the leaping blurred object) with clematis on a trellis by the Sundial Circle in the garden.

This shed was home to the Cooks for six months after the fire. It was built as a workshop where the Douglas Cook Centre is today and later shifted to the pavilion area. It was demolished in 1984. The totara garden seat is one of twelve purchased for £1 each in the early 1920s and all are still in use today.

The fire of February 1931 destroyed the Cook homestead completely. The bare slopes of Orchard Hill feature in the background.

art and many items of fine furniture. Family jewellery was also destroyed and Douglas later spent hours vainly raking through the ashes in search of the diamonds from that jewellery.

For several months the couple lived in a small shed at the opposite end of the lawn from the house site while Douglas designed a replacement home. Construction began in 1931. Money was hard to find at the time and Douglas had designed the house to be built in stages, beginning with the west end which comprised two small bedrooms, a bathroom, kitchen and lounge. Douglas's office and another bedroom were constructed later, separate from the first part but in correct position for the planned final layout of the home.

Douglas and Claire remained childless and in early 1933 adopted a six-month-old infant son who they named Sholto Douglas Cook. There was considerable excitement at Eastwoodhill at the time of the infant's arrival, and his christening was marked by the planting in Cabin Park of a blue Mount Atlas cedar, surrounded by a circle of Italian cypress.

Park planting continued. During 1934-35 Douglas took into his planting area the old horse paddock, then growing more scrub than useful grass, and planted groups of conifers which were to form the nucleus of Cabin Park. He also pursued a plan to cultivate frost-sensitive plants more successfully. – "We started the Burma Road. A zigzag path was made to the top of the hill, perhaps a climb of 150-200 feet. As six feet of sod was put over on the low side from the path we soon had ample beds to take out more tender subjects which were moved to higher levels. Now (in 1949) we grow our own sweet and cooking almonds, that most delicious newish fruit feijoa in six varieties, *Passiflora edulis*, and two varieties of the banana passionfruit *Tasconia taylorii* and *T. mixta quitensis*. We fail to flower *Luculia gratissima* because the slightest frost destroys the bloom, but such things as *Mimulus glutinosis puniceus*, *M. 'Mrs Sekoles'*, *Lippia citriodora*, *Leucospermum reflexum*, *Jacaranda ovalifolia*, all the Lantanas, most of the Australian banksias and grevilleas and most of the South African ericas do well ... The Burma Road is most useful and has enabled us to grow many shrubs much too tender for other parts of the garden." Claire accompanied Douglas on a trip to Britain in 1936, leaving Sholto in the care of Claire's sister Margaret in Auckland. Claire recalled, in the 1970s: "I bought a second-hand car so that I could take Douglas to see many of the famous gardens. We often timed our travelling so that we could see gardens on their open days, but we also went to famous gardens to which Douglas had introductions through the Royal Horticultural Society. One of these belonged to one of the Rothschilds where there were more than 30 gardeners. We also went to most of the famous nurseries and left orders with them." Those nursery visits included Douglas Cook's first direct contact with Hillier's Nursery in Winchester with which he was to have a long and close association.

Relations between the couple had become severely strained by early 1937. One historian has suggested that Douglas was not suited to marriage – a fact which he acknowledged several years later. His principal motive in becoming married may well have been to provide Eastwoodhill with a hostess. Claire began expressing her concerns to her friends at Ngatapa.

An Anglican, Claire had developed a strong friendship with Canon (later Archdeacon) A. F. Hall, Vicar of the Holy Trinity Church in Gisborne, and his wife. Neighbours recalled that Canon and Mrs Hall arrived at Eastwoodhill while Douglas and Bill were in town and helped Claire and Sholto to leave without fuss. After Claire and Sholto had returned to the family home in Auckland, Canon Hall arranged a place for Claire as

Claire with their son Sholto who was six months old when he was adopted. (Photo Jan Potter)

Sholto, aged two years and ten months, in June 1935. (Photo Jan Potter)

Sholto and Douglas Cook in Austrian outfits, souvenirs of the 1936 overseas trip. (Photo Sholto Douglas Cook)

A Cook family gathering in 1936 shows (from left) Douglas's sister Sheila and her husband Clive Wily, Felicity Wily, Claire Cook, Jan Wily, Peter Wily standing beside his uncle and godfather Douglas Cook with Sholto kneeling. Peter stayed on to help with the formation of the new entrance drive, now the present exit drive. It was a longer and less direct approach to the house and replaced the Poplar Avenue. (Photo Sholto Douglas Cook)

housekeeper to the Rev. James Young (later Archdeacon Young), who was a widower with a young family. While Claire assisted the vicar in bringing up his children in his home at Wanganui, Sholto became a loved member of the family. He was a particular friend of Peter, the youngest of James Young's boys.

Claire never saw Douglas again but they never became divorced. She eventually married Archdeacon Young after Douglas's death in 1967. She revisited Eastwoodhill in 1973 "when it was in very poor condition" and was thankful to know that its maintenance and improvement had been guaranteed. She contributed towards a fund for its upkeep. Claire Young died in Nelson aged 93 on April 1, 1988 without ever having spoken or written disparagingly about the difficulties of her life with Douglas.

Sholto Cook left for England in 1947 and spent two years at the Royal Navy Training College as an officer cadet before failing an eye test. While working on a St Albans farm to the north of London in 1952 his uncle James Cook, of Aberdeen, wrote passing on a request from Douglas for Sholto to return and help at Eastwoodhill because a manager had left. Sholto did so and for eight weeks worked until the crisis was past. He later found a job on a farm near Dannevrike and returned to Eastwoodhill just after Christmas of 1952. In June 1953 he celebrated his 21st birthday with his father.

It was about August when Sholto was thrown over a fence and down a hill by a young horse, spraining an ankle on a clump of

Seasonal views of Eastwoodhill from Semmen's Hill looking north showing The Circus, Cabin Park and Corner Park.

Summer

Marion MacKay

Compare these to the 1925 view inside the front cover.

Autumn

Marion MacKay

Winter

Marion MacKay

v

Approaching the exit gate in 1961.

W. D. Cook

Wisteria on the exit drive, Corner Park at the intersection with The Ride in 1962. Bolle's Corner, on the left side of the drive, was then covered in trees.

W. D. Cook

The seats from 1926 laid out on the House Terrace, Homestead Garden in 1962.

W. D. Cook

The newly-planted Orchard Hill from the Homestead Garden in 1961.

W. D. Cook

Glen Douglas from The Plateau in 1962.

W. D. Cook

Birch Hill in 1961 shows in the centre, looking from above Blackwater.

W. D. Cook

vii

A guided tour in progress.

Picnic tables near the car-park.

Maps and colour-coded walks allow first-time visitors to easily explore the arboretum.

rushes. Douglas was furious. Sholto recalled that his father was in a bad mood at the time.

"You're no good for the job," said Douglas. "You may as well get a job in town."

No doubt the storm would have soon passed, but Sholto took Douglas at his word, and left. He worked on an East Coast sheep station for a time, then ran a team of packhorses before moving to the Bay of Plenty where he became a fencing contractor for 28 years. He never returned to Eastwoodhill during his father's lifetime, but when Douglas was ill in the Morris Convalescent Home in the 1960s, Sholto was on a visit to Gisborne. He took his wife Dawn and their three young children, David, Claire and Sholto jnr to see him. It was an emotional moment for Douglas when he saw the toddlers for the first and only time. In 1992 Sholto returned to Eastwoodhill to take part in the official opening of the Douglas Cook Centre for Education.

Douglas Cook's life would have been far from cheerful in his large house after Claire and Sholto had left. – "I'm a lazy housewife. I don't cook much for myself… I don't say much but I'm lonely here, and won't listen to people who say I should have a housekeeper. I've had two housekeepers in my life. The first wanted to marry me, the second only to sleep with me." He had a kitchen set up at the Sanctuary end of the house which comprised a bathroom, toilet, office, bedroom and his library. Frequently his meal consisted only of boiled rice. Bill Crooks would visit for a gin after work and normally would return for a chat after his evening meal.

Douglas enjoyed his regular spot of gin with Bill and often served sherry for visitors. With his dry sense of humour, he enjoyed, as a joke, hiding the odd bottle in the park. While walking with a particular friend among the trees he would suddenly pause and dig into a marked area of the ground, or reach under a bush, to produce a bottle as if by magic. When a neighbour, Wynne Sherratt, was serving in Italy in World War 2 he received a "soldier's parcel" from Douglas. It appeared to contain only a stale loaf of bread. But it was a remarkably heavy "loaf". Inside was a bottle of whisky. Douglas, however, never became a heavy drinker. – "I've had periods of booze in my life. I never liked it. I drank because my mates did and I was lonely. Bill and I have one drink at night at 5.30 when work is

Peter Wily (left) and Sholto Douglas Cook at Eastwoodhill in 1995, the first time they were together since 1936. Peter greatly resembles Douglas Cook. (Photo Dr Susan Perry)

over for the day. It bucks me up then. But I've never been what even an enemy would call a drinker."

The Sanctuary, at the southern end of the house was Douglas Cook's private den. Self-contained for bathroom and cooking facilities, this was his evening area where he did letter writing and label punching etc. His extensive botanical and other book collections were kept in the library together with many boxes of correspondence.

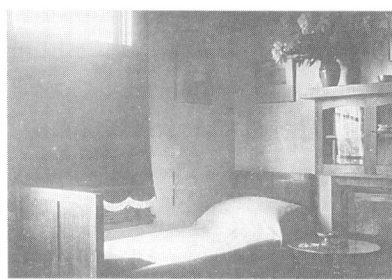

On August 25, 1937, imported material ordered during his visit to Britain began to arrive. One hundred and thirty-four different species were included in the first consignment alone. Even from New Zealand nurseries during the late 1930s there was a good range of plant material. The outbreak of World War 2 slowed down the rate of planting and cut off supplies of material from abroad. However, some planting continued.

The costs would have been horrendous by the standards of a normal farmer. In 1936 orders from Duncan and Davies of New Plymouth alone amounted to £85, equivalent to a working man's wages for half a year. Most certainly the costs would have exceeded the profits from the Eastwoodhill farm. As an old man in the 1960s Douglas was to reflect: "I came here 55 years ago. I've worked and put practically all I've earned off 1250 acres into this corner (the park). Averaging well over £1000 a year, £55,000 is a lot of money."

In fact, he also had to channel Scottish inheritance money he received over the years into the pursuit of his dream, as well as financing travel abroad. Some of the money was kept in Britain and used to purchase plants. On one occasion in the early 1960s, when Hillier's Nursery was "frozen up" by icy temperatures for two months, he drew £1000 from his London bank and lent it to the nursery at 5 per cent. – "I could have got 8 per cent!" Douglas "juggled" his wool clips in an effort to gain maximum profit from them, stockpiling a clip in the woolshed when prices were low and gambling that the price would rise - a gamble that often paid off. But he found it necessary, on several occasions, to mortgage his property for extra money.

At the end of the war in 1945 he took into his plantings the 25-acre area that was to become Douglas Park. The general public of New Zealand, or even of Gisborne, had little or no knowledge of the botanical wonder that had been unfolding at Eastwoodhill. Douglas, however, was gaining recognition among more knowledgeable circles. From 1943 Douglas was a member of the National Arboretum Committee that assisted in the establishment of the Horticultural Faculty at Massey Agricultural College. In 1948 he was elected a Fellow of the Royal New Zealand Institute of Horticulture. He became the first New Zealand member of the International Dendrology Union, proposed by Lord Digby. Further honours were to follow.

Hard on the heavy heels of World War 2 came a threat that was to haunt Douglas Cook and spur him to almost frantic efforts to plant all he could of Northern Hemisphere plant

This spring view of the Homestead Garden was taken in the early 1920s from the site of the present pavilion. It shows the beds of tulips purchased from Holland in 1920.

species within his lifetime. The arrival of the Cold War, with nuclear powers as the potential adversaries, convinced him that there would be a conflagration that could destroy much of the plant life of Europe. His dream for the park developed international significance as he began to think of Eastwoodhill as a future repository of plant material for the gardens of Europe following the devastation of war.

In a letter to Harold Hillier in October, 1950 he said he was afraid to leave New Zealand lest an outbreak of war prevented his return to his park planting. In fact, he made one final trip to Britain and the Continent in 1954. The pessimism there about the possibility of peaceful solutions between the powers, and talk of missiles with atomic warheads, did nothing to allay his fears. As the years passed he dreaded the possible effects of nuclear devastation on plant life in the Northern Hemisphere. He was a true "environmentalist" long before the term became fashionable.

Roses and herbaceous plants bloom in the Homestead Garden in this late-1930s view looking down from the Palm Terrace. The bare slopes of Orchard Hill in the background. A manuka brush fence gives some shelter from the north-west winds. The long path was a feature until recent redevelopment of the garden.

5 A Home for Rhodos

Douglas Cook had a passion for rhododendrons, or "rhodos" as he frequently called them. The fact that they did not fare well at Eastwoodhill was a disappointment, but he never stopped trying. Indeed, by 1949 he was growing over 110 species and between 250 and 300 hybrids. He lamented at that time: "Ours is not the ideal climate for rhododendrons but it is near perfect for the human race. I am sure we can grow rhododendrons here eventually but it may take time to find the right situations. I think the ideal will probably be found in an open southerly situation with high distant shade to the sunny north. I believe here that protection from drying winds is more important than shade from the sun."

No aspect of his plantings demonstrated more vividly his astonishing perseverance in all gardening. As early as the 1930s, after one planting project, he found that the rhododendrons wanted to be moved. – "It was really heavy work. I moved most of them on my back, but some had to be moved on a stretcher contrivance." In 1950 he wrote to Harold Hillier: "No more rhododendrons. Ours is a delightful climate at Eastwoodhill but

Development of the Homestead Garden began in 1920. Black pumice soil was dug from the paddocks to build beds for the collection of tulips imported from Holland. In the background are the bare slopes of Rock Ridge (right) and Lookout Hill above The Highway (left).

not for rhodos." That ban proved brief. He continued to plant them and by January, 1963 he commented about plants arriving from England: "When the rhodos come there will still be another lot, probably 60 to 80 rhododendron cuttings... By that time our already overworked shadehouse space will be bursting at the seams."

The combination of Douglas Cook's frustrations in growing rhodos, and his love of Taranaki and its mountain was to prove an auspicious one for the nation. He knew that rhododendrons flourished at Pukekura Park in New Plymouth. – "Quietly for 14 years I kept my eyes open for the perfect home for them and eventually, in March 1950, with the help of others, found what I considered the ideal spot. A section of forest land from which the big timber trees had been removed, a rainfall of 150 inches a year well spread over the 12 months, a height of from 800 to 1600 feet above sea level. Streams, gullies, shade, open spaces and just about every condition most rhododendrons would enjoy. With filmy ferns happy on tree trunks even epiphytes should grow on trees."

The Lodge at Pukeiti after its extension in 1962. Douglas Cook had undertaken to get a flag from every country in which Pukeiti had a member. Of these 12 flying, he had acquired nine. (Photo Douglas Elliott)

Douglas, with his friends, had found his rhodo site on the lower slopes of the mountain – 153 acres including the 1601ft peak called Pukeiti, Maori for "little hill". The view from the top of the hill deeply moved him. – "I saw it at 3.00 in the afternoon and I found the owner and paid a third down before 8 p.m. that evening. Never a regret. " He offered the site to the New Zealand Rhododendron Association at its annual meeting in October, but the organisation declined his offer through lack of finance. – "I announced that Pukeiti would still go on as a Rhododendron Park." He decided to offer the land to a group of no less than 20 members who would contribute £50 a year for not less than five years. With the help of a fellow plantsman, Russell Matthews, he succeeded in finding keen rhododendron lovers who were prepared to make a continuing, annual contribution. The Pukeiti Rhododendron Trust, to which Douglas gifted the land, was formed in October, 1951. A group of 40 gathered at Pukeiti a few days later and Douglas cut the ribbon at the entrance to the hill (now called the Cook Entrance). The intrepid band walked three-quarters of a mile in driving rain and through mist to the summit.

Douglas firmly set the direction of the trust: "I insisted it must be a private trust and that no donations be accepted from government, county council, city council or any such public body. We want no dictation in our affair. We are our own bosses and we alone say what is to be done." He also pursued membership donations among the nobility and fellow-members of the International Dendrology Union during his visit to Britain in 1954. – "At Brodick Castle on the Isle of Aaran, Lady Elphinstone (sister of the Queen Mother) introduced me to the Duchess of Montrose. Both joined Pukeiti and the Duchess of Montrose gave Pukeiti their Scotch flag. Lord Bledisloe I called on and spent a morning with him and Lady Bledisloe and he gave Pukeiti a Union Jack. I undertook to get the flag of every country in which we (Pukeiti) have a member and I got nine out of the 11. We have 11 flag-poles at Pukeiti along the stone wall and we fly them on public days. A gay scene.

"I got about 20 titled people in Britain as members of Pukeiti. Lord Glasgow died last Christmas. Ninety-two years I think and he was a member. In a recent letter Lady Glasgow tells me she is going to keep his Pukeiti sub going."

The building of The Lodge in 1954, the opening of bush walks, access to streams, a waterwheel, terraces and the entrance gates were all developments at Pukeiti in which Douglas was able to enjoy shared pride. He became a council member of the New Zealand Rhododendron Association and was elected patron of Pukeiti.

On public open days Douglas was always immaculately dressed. Here he chats with a visitor in the sundial circle. By this time of his life, he and Bill were unable to keep up the work and the Homestead Garden had lost its herbaceous element and was reduced to the shrubs and trees.

Part of Pear Park in 1939. The small valley was called 'Below Bill's' and was planted in trees in 1926 and includes pin oak, red oak, scarlet oak and American elms. Bill Crook's cottage was later to be erected on the extreme right after the felling of the pines and levelling of the site. The manager's cottage in the centre was called the Cliff Cottage as it was built on a spur, though today the presence of so many trees apparently smoothes out the landscape. This cottage was moved for Adrian Cave in 1952 when he purchased the larger part of Eastwoodhill.

Besides rhodos, Douglas Cook had a fondness for daffodils dating from childhood. He wrote frequently about them. In 1950: "In the spring which seems to start in winter (if that isn't too Irish) there are hundreds of thousands of daffodils to begin with; then come the vast majority of flowering trees and shrubs." In 1964: "I wish you could see our Daffodil Paddock, it's a sheet of colour, yellow and white. Millions! These are all common old ones 50 years old. I bought one of each, ld up to 6d, when I was at school. These are all their children." Perhaps the real beginning of Eastwoodhill as a form of public park was a decision by Douglas to establish Daffodil Sunday once a year. He allowed each visitor to pick as many blooms as he or she could hold in one hand, but no blooms from the garden near the house. A friend recalled an incident when a woman had her arms full of blooms and was still picking.

"That's more than enough," Douglas snapped. "Get off my property!"

He also initiated open days in which he allowed various organisations, among them the Jaycees, the Youth Hostel Association, the Save the Children Fund and the local branch of the National Party to charge at the gate in order to raise funds.

Douglas provided the water for "bring-your-own-cup" afternoon teas, usually served from the House Terrace. Hundreds of people attended. He wrote: "Last Sunday the place was open to the public and we had a thousand people and a swarm of kids. The Youth Hostel Association I lent it to made over £100 out of the day." He commented that there was seldom any damage, never any litter to clean up, and no plants were ever stolen. – "Occasionally a child will collect a lead name tag to melt down to make a sinker for his fishing line. I have seen English picnic spots at the end of a day and I am glad that our public here are more tidy-minded."

His greatest joy, though, was to escort plantsmen through his park: "It is not a one-day walk for anyone who is interested to see this place now, and I fear I get little pleasure from showing people round who are not. There are many who come here and view a sight and then go away and air their ignorance - but I am pleased to show true gardeners round or people who appreciate beauty and can sense the restfulness of the place."

In 1952, the importation of plant material continued apace, Douglas Cook was 71 years old and had dedicated his remaining years to the development of his garden and so he sold 925 acres of his farmland to Adrian Cave for £20,000. Some of that money went into development of the Circus area in Eastwoodhill; some was used to develop ponds throughout the park. Adrian respected plants and admired Douglas's dedication to his planting. The two men seemed to have a good relationship. But that was tested during a drought in 1957-58.

"There was quite a lot of grass on his side of the fence," Adrian recalled.

He obviously knew Douglas's weakness for home-cooked goodies because he asked his wife Mary to cook a raisin cake for a visit to Eastwoodhill. There had been a long family association. Adrian remembered visiting the homestead as a child with his mother, Joan Cave, who was a keen gardener. He had been enchanted then with a whisky decanter in the shape of a dog. The tail formed the handle and the pourer was the dog's mouth. Douglas had been generous in those days with seedlings when the Caves were planting on the farm.

On this dry, summer afternoon, over a cup of tea, Adrian broached the subject of the grass that could have provided extra grazing. Douglas expressed concern about cattle. Adrian gave assurances that he would not allow them to break through the fences.

"He was adamant that he would not release the grazing," Adrian recalled, "but he enjoyed the cake anyway."

A few days later Douglas offered the grazing to another neighbour.

Adrian seldom saw Douglas in subsequent years – "As he grew older he became obsessed with documenting and tagging all his trees." He held no grudge over the strange affair of the grazing and commented later: "Few people appreciated the amount of work that went into the planting of those trees at

Circus Development

A ridge on the northern side of the Circus paddock was planted in radiata pine and was milled by J.T. Hill in the late 1940s. It was worth little in the way of return to Douglas Cook as it was prior to the development of the tanalising timber treatment process.

The earthwork development of the Circus was done in the mid-1950s. The wall of Hillier's pond shows (above and below) with the backdrop of the Lombardy poplars of the Poplar Avenue, today's entrance drive.

The Lombardy poplars (below) date from 1918 at Sherratt's pond. They were left during the 1950s alterations to the Circus and some still stand in 1997.

Eastwoodhill – the way they carried black soil to get the trees started. It involved an enormous amount of manual work."

Douglas Cook was known through his writings for his biting wit, and any "petty officials" who interfered with the smooth transportation and delivery of his plant material were prime targets. He wrote, in a 1963 New Year letter, of a customs official: "One of those government officials whose mother had no milk, so she fed him on red tape from birth. " That referred to an incident in which the customs man was querying the uplift of five *Magnolia macrophylla* because Douglas had not yet received his permit. The request for the permit had been delayed in Wellington, presumably by another "petty official".

"Personally I'd like to know what bloody right has any government department to say whether I may or may not spend money that has been in London almost 10 years," Douglas added. "Have we no rights as citizens or are we a pack of children under office boys' thumbs. Where man's laws conflict with God's laws, I follow God's laws. Where my common sense conflicts with tuppenny halfpenny little boys' rules, I follow common sense ! " His wrath was understandable. He was going to extraordinary pains to import costly and sometimes rare plant material, only to find on numerous occasions that there were delays at wharves or at rail sidings. – "Once a case of plants lay on a Wellington wharf for three months till a shedhand very decently wrote and told me it was there and I at once wired the Express Company. 'Express Company' mind you ! They were quite annoyed, I think, at my finding out. It's a lot of fun importing but it has turned what hair I have white. You know, some people have to have a lot of courage to live."

On one occasion plants that were supposed to travel in the ship's coolers were put into the freezers by mistake. On another, Douglas and Bill Crooks had to drive north to Taneatua, a rail terminal in the Bay of Plenty, to rescue a box of Taiwanese rhododendrons that had been left there, sitting in the sun, through the Christmas holidays. Douglas related another saga of rhododendrons which arrived in "cool vegetable chambers" aboard the Paparoa, which berthed at Auckland on January 14, 1963: "Sounds all right but my experience of Auckland Harbour in December and January is six to 12 ocean liners lying at anchor waiting for berths while dockers have sit-down strikes and morning and afternoon teas lasting most of the month.

"Getting things by ship about drives me mad. When eventually they do land, the Express Company are too busy to forward them for at least a week, then they put them on a train for Taneatua where they arrive and sit another week with temperatures 85 to 87 degrees F because Gisborne is too thirsty and out of beer, and beer must go through first on service lorries. Eventually they reach Gisborne Railway Depot and they just expect I'll know they're there and don't bother to ring me. After a week they decide, 'better get these darned things out of our road' so they ring Eastwoodhill 30479 – No reply! The poor dears don't know Eastwoodhill is now on the Ngatapa exchange and they have to ring Ngatapa 597. Eventually, with winter almost on us, we'll get our rhododendrons grown in their cases. Long white shoots which when they're unpacked say, 'to hell with this country. I've had it!"

It must have taken huge volumes of patience from a notoriously impatient man to continue his importing of plants.

One of Brian Sherriff's (Wade Studios) early 1950s aerial shots. On the right adjacent to the Wharekopae Road is the clearing of The Circus paddock. Douglas Cook's woodlots of radiata pine were a rare feature until the government encouragement schemes in the 1990s. The homestead features in the upper left of the planted area. Neighbour Wynne Sherratt's future home site features at lower right. Part of Glen Douglas is at the lower left corner, the high point being Bishop's View of today, with the roof of the Cabin showing. The open lower area is now part of the native plantings of Gondwana Bush. On the extreme left side is the Douglas Cook Walk leading to Douglas Park from the Orchard Hill tank.

Early Days in The Homestead Garden

The sundial circle has featured irises, dahlias as well as many other planting styles and is nowadays in lawn. Four brick steps mark the points of the compass.

The brick work in the garden was done by Percy Beale in about 1936. The Palm Terrace here featured a formal line of seats and Chusan fan palms, the latter "a mistake" admitted Douglas Cook.

Looking down from the Palm Terrace in the late 1930s. Already tulip and herbaceous beds are replaced in part by a lawn.

Full collections of herbaceous peonies were imported, shown here in the late 1930s and a number of cultivars still remain today.

A bachelor path led from the back door to the butter-cooler, a cupboard made of a pipe buried in the bank. The path was less than three feet wide so was a one-person path when the plants grew.

A view of the garden from Orchard Hill in 1931. Compare this with the 1910 view on page 10

6   Jungle by the Sea

Douglas Cook liked to keep himself up to date with the business "doings" of Gisborne. He made a habit when in town of calling into the office of Ball and Crawshaw, real estate agents, to chat over a cup of tea. During one call in 1956 a young Brian Crawshaw (later to become Mayor of Gisborne and a member of the Eastwoodhill Trust Board) told him of a block of land for sale at Wainui, a seaside village 5km along the coast from town. Douglas bought it and in January, 1957 arranged to have a cottage built on the part of the property fronting on to Murphy Road.

He called the place The Jungle and it was to become a source of interest to him in the difficult years of the 60s. The property offered him a frost-free environment for propagation and a place to grow tropical plants, including a banana tree, that would not have survived at Ngatapa. As well as an opportunity to grow tender plant material he was able to hold imported orders and "grow them on" before moving them in a couple of years to Eastwoodhill.

Douglas had chosen, as builder of his cottage, Barrie Currie who lived next door to the site at 35 Murphy Road. Barrie went to Eastwoodhill to receive his instructions and was told he was to use only rimu, no pine. He arranged deliveries from Aicken's Timber Yard in Gisborne. On one occasion he needed a small quantity of timber for a job on his own property and arranged for it to be delivered in the same truckload as some of Douglas's timber. Thus the cost of delivery would be shared, with a saving to both men. Douglas, obviously misunderstanding Barrie's motives, rang the carrier and cancelled the order. It soured their relationship. Barrie completed the cottage but, though they were next-door neighbours, they never visited one another.

Douglas did make some friends at Wainui. Mike and Patricia Jeory met him at the local store. Conversation flowed easily: Mike was executive officer in Gisborne of the Ministry of Agriculture and Fisheries and was knowledgeable about agriculture and horticulture; Patricia was a journalist who enjoyed gardening. On one occasion Douglas bemoaned the fact that he was no longer able to buy the "nice little cakes" which were one of his passions. Patricia baked him some and their friendship was sealed. The Jeorys were looking for shelterbelt plants and were able to get advice from Douglas; he gave them seed to grow a "snail plant".

Mike and Patricia invited Douglas to meals on several occasions. They visited The Jungle only once, but frequently drove to Eastwoodhill and took their children for walks around the grounds. On most visits Patricia would take a cake, or a tin of those "nice little cakes". In a letter to them from Eastwoodhill in 1962

Douglas Cook and Bill Crooks walking along Gladstone Road in Gisborne. A man and a boy is how Douglas Cook referred to himself and Bill, especially in the context of work in the garden. The height difference makes it look so. Douglas' suit looks a little small - he was known not to spend greatly on clothes preferring to buy trees.

The new cottage at Wainui that Douglas called The Jungle. It was the area behind the house that warranted the name.

The path went between the shed and the house through a gate into the main part of the garden that was surrounded by an eight-foot iron fence. There he could work unseen by the neighbours. Douglas jokingly planted a fig at the gate 'for the shy ones' to pick a leaf off if they felt he needed covering.

there was a strong hint that he would appreciate some more: "Thanks very much for the cake. I'll enjoy that I assure you. Bill and I use a lot of biscuits at morning and p.m. tea so I started getting very big tins of broken biscuits. About a quarter were whole ones." He went on to grumble about the high price of food including that for the biscuits he used to enjoy in his school days.

The wide-ranging letter also reported on the scene at Eastwoodhill that winter:

"Had our first frost of the season and my hands are so cold I can hardly write. It's amazing to be only one week from the shortest day and this one first frost. There has been a riot of camellia blooms and lasiandra in front of the house; also ginger – so sweet-scented, still in flower. Well, it had to come and I'm grateful to God for letting winter come so slowly and gently. It's been a cranky season, autumn colour was erratic and generally poor. The poorest season for many years but, poor dears, the plants just didn't know whether to dress for summer or winter, nor when to have their children. We had masses of camellias in April. Of course, far too early. This morning all our exposed camellias will be cut but inside the bushes plenty still.

"The dams you saw in the making are full and both running over. A great improvement on the scene. One looks like whitewash and I think both will take two to three years to clear. I must get their water tested to see whether acid or alkaline. I hope acid but I doubt it for a few years.

"I have started making a list of every tree and shrub growing at Eastwoodhill 1962. Taking it carefully section by section. I'll get it finished in fine weather then on cold wet winter days make it alphabetically correct. From then on we order only new material except where wanting mass for landscape work. I've got nothing coming from England this season but when my list is completed may order a few new and rare things.

"At the moment we are working on the first section of a two-year clean-up programme. We've started up around the Cabin and the weather has let us get a lot done. It's so terribly overgrown that there's a tremendous lot to do and whole rows of trees to cut out entirely and burn." In the section of his letter about prices he went on to complain about the economic situation in general:

"Factory butter was 1/- per lb, now 2/-. That increase is just madness. The Government are subsidising the consumer and at the same time women are mean enough to be using all sorts of greases in place of butter which here is the cheapest in the world. It's a mad period we're living in but a day of reckoning is coming. Wages must topple and so must prices of goods in shops. More

people must work instead of loafing their way through life. The farmers are not moaning but they've every reason to. Their produce is three times or twice as profitable as it was 15 years ago but wages are seven times what they were."

Patricia was to remember Douglas as "a rather lonely, sad old man". She was intrigued, on Eastwoodhill visits, that the large dining-table was always set, complete with candelabra and the best crystal and silver, as if he were permanently expecting dinner guests – "There was something almost eerie about it that reminded me of Great Expectations." When Mike and Patricia had meals with Douglas they always ate in the kitchen.

As a lad Robin Crooks, a son of Bill Crooks, used to help his father with the gardening at Eastwoodhill and when The Jungle was created he often travelled to Wainui with Bill and Douglas at weekends or during school holidays to assist there. One of his heavier projects was digging a large plot on the land behind The Jungle and planting it in potatoes. He recalled that many plants "just went berserk" in the frost-free conditions at Wainui – "I spent a lot of time trimming them back. Plants that might be a metre tall at Eastwoodhill would be two metres at Wainui." Douglas did not thank Robin for his labours, nor pay him. But years later he gave the young man one of the back sections – "That was just his way. It showed that he appreciated my efforts but chose not to tell me at the time." When Robin married, he and his wife lived for a time in the cottage and became friendly with their neighbours, Barrie and Beryl Currie.

The mahogany table with Douglas Cook's Georgian silver and his own design of crystal, a Waterford pattern which he had cut into a Stuart stem. On one occasion, soon after his and Claire's return from Britain in 1936, they entertained neighbours to a midday meal which featured fish. Much had been packed away but Douglas refused to let the meal start until the fish knives were found, by which time the meal was cold.

Of one tenant in the cottage Douglas wrote: "Yesterday my tenant at The Jungle called and paid the rent. In fact, instead of paying four weeks in advance he paid eight weeks to 23rd February as he'll be busy at wool sales. I asked him if he had plenty of money as now, while wool is low, is the time to buy. He said he was all right but had to be careful. I offered him some and he offered me 10 per cent. That's very attractive. Wish I had a lot to lend him at 10 per cent. I have to be very careful but my needs are few."

In the 1960s he once considered selling the property – "I offered my beach cottage and tropical garden, The Jungle, at £4000 but the lady turned it down. O.K. If she asks again it will be £4500." At the same time, concerned about his future, he was conscious that the cottage might provide him with a home closer than Ngatapa to medical aid if he became seriously ill.

There is a postscript to the story of The Jungle: In 1992 a Gisborne artist, Chris Morrell, and her husband John acquired the property. Chris turned the shed at the back into her studio which she called The Batch (spelt correctly with a "t" because it referred to her collection of paintings). Chris created a cottage garden and

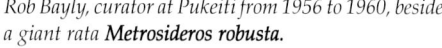

*Rob Bayly, curator at Pukeiti from 1956 to 1960, beside a giant rata **Metrosideros robusta**.*

Pillar apple Malus tschonoskii *at Birch Hill Pond, Douglas Park in 1994. It has since died.*

Stephen Jones

Swamp cypress Taxodium distichum *at Rock Point Pond, Douglas Park.*

Teresa Campbell

ix

Formation of the Birch Hill Pond, Douglas Park in 1961.

W. D. Cook

Birch Hill Pond in 1963.

W. D. Cook

Rock Point, Douglas Park and the Douglas Cook Walk in 1959. The pond was formed in the valley in the foreground in about 1960. Blackwater Pond is at upper left.

W. D. Cook

x

W. D. Cook

W. D. Cook

Rock Point Pond in its early days, 1961-63.

W. D. Cook

xi

Stephen Jones

The European white oak, Quercus pubescens *at Rock Point Pond.*

Kerry Fox

Scarlet oak, Quercus coccinea *at Rock Point Pond.*

Japanese maple Acer palmatum 'Osakazuki' *at Birch Hill Pond. (see similar view 1963 page x)*

Stephen Jones

The Rock Point and Birch Hill Ponds today.

Scarlet oak at Rock Point Pond.

Kerry Fox

A view of Glen Douglas in 1961 showing ponds and tracks with the trees only recently planted. Douglas Cook did indeed survey off 25 acres for Rob Bayly. Sites for two houses, a glasshouse and open ground nursery were prepared but his dream came to nought as he had no money to add to the venture. The Baylys had spent all their savings and left Eastwoodhill shattered.

planted sunflowers to use as still life models for her works of art. She also painted images of many of the plants that had been established by Douglas including a gardenia and a frangipani, using the established trees as backdrops. A special model was a flowering cactus planted by Douglas. She named the work The Cacti in Doug's Garden. (Douglas would not have approved of that name. "Doug"!)

While The Jungle was a Douglas Cook success story, he embarked on another project out of concern for the future that turned into a disaster. He conceived the idea of inviting Rob Bayly, then curator at Pukeiti, to create a nursery at Eastwoodhill. Douglas was to give Rob the 25 acres of Glen Douglas and a site for him and his wife Claire to build a home. Douglas wanted the material in his plant collection to be more widely available. He also prepared a house site close by where he was to erect a fire-proof structure for the protection of his antiques, books and records. Central to the scheme was the idea that Rob and Claire would be able to look after him in his old age.

Douglas visited the Baylys at Pukeiti and stayed overnight with them to discuss the venture – "He was a perfect gentleman and he promised the earth," Rob recalled. The Bayly family, which included three young children, Jenny, Irene and Paul, moved into the Eastwoodhill homestead, in January, 1961. The Baylys occupied one end of the house; Douglas had his "sanctuary" at the other end. Trouble began almost immediately: Douglas had little patience with small children. He had once written to the Jeorys: "Two things I can't do with here – stiletto heels and small children. Sad isn't it?" It was impossible for Rob and Claire to keep the children quiet enough to satisfy Douglas. "There were hassles over the children," Rob said later. "They couldn't make a squeak without him complaining. It created a lot of tension."

Problems arose outdoors too. Rob repaired the glasshouse left from the previous nursery and began to propagate with Claire helping with the cuttings, but the water pump broke down after a fortnight. The Baylys had to pay for a new motor. There were other expenses, such as metal for the road to the nursery site, which came out of their life savings. Said Rob: "I got the bill for any repairs. I helped a lot with Eastwoodhill work but was never paid any wages. " By August of that year their savings were virtually gone and they had no legal title to the land on which to build their house. They departed, and it took the Baylys 10 years of hard toil, and a lot of assistance from friends, before they were able to establish a nursery at Wainui which was to become known as Bayly's Greenworld.

Depressed by the collapse of the Bayly plan, which would have taken care of his old age and, through the presence of Rob, provided continuity at the arboretum, Douglas became agitated about the future of his park. He was prepared to give it away to an organisation that could care for it, and preserve it for the nation. Douglas poured out his anguish in letters to his friends. One of those acquaintances was J.A.S. Howard, of Marton, who wrote to the Dominion Secretary of the Royal New Zealand Institute of Horticulture in February, 1964:

"I am writing to you just because I feel very strongly that Eastwoodhill must be preserved for posterity and yours seems to be the institution most suited to meet the challenge . . . Undoubtedly the maintenance charges for a start would be very heavy, but as the newly-planted trees grow up and form a canopy it will largely look after itself. As time goes on, and trees now six and eight feet tall approach the 60 and 80-foot height, the value of the park will be obvious to all keen gardeners."

In a lengthy and impassioned plea he added: "You will realise that I feel very strongly that you should do your utmost to see that the opportunity of gaining this magnificent national asset, which I am sure in time it will become, is not dropped or passed by. Further, I feel that the good your institute has done up to date would be cancelled if you fail in this matter."

Douglas weighed in with his own letter to the institute suggesting that its national headquarters "should own and manage the place, and Gisborne branch see to its orders being carried out".

The institute considered the letters at a meeting of its Dominion Council in March and set up a subcommittee "to give

careful consideration to the proposals". As a result, a number of experts visited the arboretum, among them Jack Goodwin, superintendent of Parks and Reserves at New Plymouth, whose detailed report concluded: "I am confident that Eastwoodhill, if preserved, will become one of the great arboreta of the world." Letters continued back and forth. There were legal as well as financial complexities and Colonel Reg Gambrill (whom Douglas described as "my senior brainy in trust matters") became involved. Some institute members felt that funds from the national lottery of the time, the Golden Kiwi, would offer a solution.

But time was running out for Douglas…

Glen Douglas from Bishop's View in 1961 with only a few trees planted, including the pin oaks in Oak Crescent in the right foreground. When the Glen Douglas dams were being formed in 1958, Douglas Cook unnerved the bulldozer driver with his unclothed presence. To the left the 'Devil's elbow' section of the Wharekopae Road was cut off by a road deviation just prior to Douglas selling Eastwoodhill. It was given to the Cook County Council as a picnic reserve by H.B. Williams to honour an offer made by Douglas Cook. The shearer's quarters adjacent to the bend in the road were used by Eastwoodhill, for the six years from 1991, as accommodation for visiting student groups.

Cabin Park

Having filled Corner Park, Douglas Cook started planting Cabin Park in 1936. It was previously a horse paddock but grew bracken better than grass. It was an exciting time for him as he was planting material new to New Zealand. On the hill behind the cabin he tried many tender plants some of which proved to be hardier than thought. Cabin Park was where the first material imported from Hilliers in 1937 was planted. A track on the southern side, today called Shady Way, then bore the unusual name of 'The long road to Poland'.

A 1954 view over the young Beechwood, the pines in the centre are The Vista of today. To the left of the pines is where the H.B. Williams Walk and the Miniature Conifers are today. The newly-made Orchard Hill tank (centre left) was built on site in 1939 by Mr Otway when the Totangi Dome oil well was being drilled in the hills opposite Eastwoodhill. The view is from near the Patula pines on The Highway, a prominent look-out point on the mauve walk today.

A Dutch couple lived in the cabin for a while in the early 1940s. The husband, Henk, worked in the garden. He is pictured here, like a giant alongside Bill Crooks, on one of the newly-dug paths above the cabin.

The 'Canadian Cabin', with fibrolite exterior and match-lined rimu interior was host to many of Douglas Cook's naturist friends over the years.It also boasted a well-stocked sherry cupboard for his garden visitors. Nowadays it hosts only bees in the ceiling. The cabin was also the centre of an armed hold-up in the mid-1940s. Dick McGill, a Social Welfare placement in his mid-teens boarding with neighbour Dixon Jobson, armed himself with a .303 rifle stolen from neighbour Jimmy Fitzgerald and fired at Mr Jobson and neighbour Ralph Warner late one afternoon. He broke into the cabin and there followed a long siege the next day with further shots at Constable Allen from Patutahi, who weighed in at 20 stone and was unable to hide his bulk behind the ten-year-old trees. Later that afternoon, McGill surrendered to Detectives Waterson and Sneddon. He received a jail sentence but a couple of years later was involved in another incident at Poroporo near Whakatane where he shot and fatally wounded traffic officer John Kehoe. He committed suicide shortly afterwards.

7   A New Pair of Slippers

Alone in his home at Eastwoodhill, surrounded at night only by darkness and by silence, 80-year-old Douglas Cook was in a state of despair. Often unable to sleep, he would rise at one or two o'clock in the morning and pour out his fears on paper in letters to his friend Graham Weston, then research liaison officer at the Forest Research Institute, Rotorua. Those fears were not only for his own future but also for the future of Eastwoodhill. It was now 1965. For Douglas, time had dragged painfully since the Royal New Zealand Institute of Horticulture had begun to examine the possibility of forming a trust to administer and preserve the arboretum. The frequency of his letters to Graham, sometimes only one or two days apart, reflected the depth of his anxiety.

Graham had first "discovered" Eastwoodhill through an

H.B. Williams did not purchase Eastwoodhill until June 1965 but was involved earlier. Here Douglas Cook, in the same Austrian hat as seen in Chapter 4 poses with others on the Homestead Terrace. On the back of this photograph in Bill Way's handwriting is the following inscription.

"The men who came to Gisborne in April 1964 at request of Colonel R. Gambrill (Nolan & Skeet) to prepare report on value of Eastwoodhill for H.B. Williams. Left to right J.D. Salinger (Horticulture Dept. Massey University), J.W. Goodwin, V.M.H. (Director of Parks & Reserves, New Plymouth), W.D. Cook V.M.H., J. Yeates, (Botany Dept. Massey University), J. Living of the R.N.Z.I.H."

The V.M.H. (Veitch Medal of Honour) signifies that the person has been awarded the Veitch Memorial Medal by the Royal Horticultural Society.

Douglas Cook admires a Knap Hill azalea in the mid-1960s. In these declining years of his life he found it difficult to get about his plantings. Bill Crooks kept the main tracks open by mowing them and would drive 'the boss' about in the car. By this means Douglas continued to enjoy the rewards of his life's effort. (Photo Brian Sherriff, Wade Studios)

institute colleague, Bob Burstall, who was roaming through country districts looking for large, old trees in order to measure their heights and volumes as part of a botanical study. This led to visits to Eastwoodhill – the first for Graham in June, 1964 when he met Douglas Cook. (Many years later, from April 1980 to May 1986, Graham Weston was to represent the Department of Scientific and Industrial Research on the Advisory Committee of the Eastwoodhill Trust Board.)

On February 16, 1965 Douglas wrote to him: "I'm very depressed this morning. I'm worried over the future of Eastwoodhill – naturally, my life's work. I want to see it enjoyed by future generations of New Zealanders but I'm depressed and worried about the lack of constructive interest taken in it by the institute. I refuse to admit my life has been a failure. I've worked (and spent) hard to get every beautiful tree and shrub which will grow here into New Zealand... All I can say is that their discouragement is not in the interests of present and future New Zealanders. Chicken-hearted doubters get nowhere. One has to have enthusiasm and ability to be a success in whatever one takes up... "

His state of health cropped up frequently in his writing:

"I was in town yesterday morning but felt so ill I had to come home. I'm old and I can't think. I need someone to look after me (and affairs) and there is no one. I never thought I'd be lonely at Eastwoodhill but I am. I can't work and I try not to brood over being incapable..."

"I'm sorry Graham but I just can't get well. I can't pick up where I was when I went into hospital..."

"If I walk too far in the park I can hardly drag myself home. Old age has caught up with me. 80 and 7 months. Well, at least I haven't loafed my way through life. I've worked and done the best I could with a poor empty head..."

"I'm constantly reminded by my heart that I may go anytime. It doesn't worry me at all. I've done my best by my country. The rest is up to the people themselves."

There was one touch, poetic in its pathos, after autumn: "I'm sorry you missed this year's autumn colour. Graham, I feel my trees have been giving me a happy send-off and saying goodbye." Plants, of course, were still constantly in his mind. He wrote that the English lilacs were in flower and sweetness "and the wee babies are bursting their buds. I keep telling them Hurry Hurry Hurry as winter is near."

Some letters told of the sale of his stamp collection (£540), a Persian rug and some of his silver (£5 for each teaspoon) – "I've got to sell things now to pay Bill Crooks and to eat". But even with that news he interspersed items and questions about plants. He asked Graham, "Have you anything in the tree line in Rotorua you think I should get?"

Despite his frustrations and his sense of foreboding, he continued to plan new plantings. He had written in 1964 to the Arnold Arboretum in Boston, who referred him to John C. Webster of Swarthmore College as "America's expert on lilacs". He now had a list of varieties he wished to obtain. He told Graham: "I have a collection of lilacs arriving from England in Wellington tomorrow, others coming by air. I may be getting six to eight from Manitoba in Canada where they'll be frozen into the soil till April. Tough guys. " He asked Graham: "What I can't remember is, do sheep eat lilacs? Have you any idea? I don't think they do." In May he reported: "The 57 lilacs new to New Zealand arrived some weeks ago and I've crocus by the thousand coming from Holland… Even if I die tomorrow, I'm still getting in all trees and shrubs worthwhile." On one hand he felt that he could not carry on and was tormented by the fact that the arboretum was "slipping back". On the other, he declared that he was "all out" to get everything in trees and shrubs into New Zealand before the next world war – "We'll all be too poor after it, for long years." This was at the time of the Cold War, with nations stockpiling nuclear weapons.

Besides the plants there were other diversions to lighten the tone of his letters – the house cow for instance: "I'm all dressed up for town but Bill left a gate open and the b——— of a cow has taken it into her head to have a look how big the park is and it's close on 10 a.m. and she can't be found. She may be feet up in a dam somewhere. I hope not. She's pretty brainless though very quiet. She had a calf four years ago and has just gone on milking splendidly ever since. Her mother didn't tell her she should have a calf every year." Many years later Molly Scrivener, who as a keen member of the International Dendrology Society volunteered time to assist with naming plants at Eastwoodhill, was to remember Bill Crooks telling her that Douglas actually chased him with an axe on that occasion!

He wrote, too, about beauty and the arts. Reminiscing about his visit to Bavaria, the Tyrol and Switzerland 10 years before and about the wonderful beauty, he mused: "I'm not religious but I feel near to God and nearer to being (or wanting to be) good in great beauty than anywhere else – great beauty and beautiful music. Modern sculpture and paintings are an insult to God. Some people have twisted minds… "

Despite his comments about his failing memory and his

Col. Reg Gambrill of Nolan & Skeet who acted as lawyer for both H.B. Williams and Douglas Cook. (Photo Nolan & Skeet)

This photo in 1960 shows (at right) the 24-year-old plantings in Cabin Park. The row of radiata pines (centre) is The Vista of today. Douglas Cook Walk (centre) heads towards Rock Point and into what is now called Douglas Park. Douglas Cook called this area Nudist Park. He flattened five sites to construct cabins for his naturist friends to enjoy the seclusion, but they were never built. Only the cabin in Cabin Park served this purpose. Of the ponds only Blackwater (left) was built at this time. (Photo Brian Sherriff Wade Studios)

"poor empty head", Douglas was constantly exercising his mind about possible alternatives to any arrangement with the RNZIH. Above all, he was determined that the future of the arboretum must be secured before his death – "My object in life is to save Eastwoodhill's life and secure it for the next 50 years; surely no one after that would be idiot enough to destroy it."

He was fast running out of patience with the institute. His lawyer, Colonel Reg Gambrill, had told him that "these things take time". Douglas's reaction: "I guess they've had about all the time I'll allow. The institute told me not to be in a hurry or do anything rash on Dec 31 (Date I said I'd withdraw my offer). Hell ! Hurry?" Apart from what he saw as time-wasting, he did not approve of the institute's reliance on being granted money from the Golden Kiwi Lottery in order to establish a trust. He called it "crawling" for Kiwi money. "I hate that," he wrote. "It's a poor show if, given Eastwoodhill, they can't collect enough privately to run it. I want to see enough income from investment to pay wages and running expenses."

Based on his experience with Pukeiti, he believed that there was an opportunity for public subscription: "If the RNZIH had tried to raise money we could have toured New Zealand for members to join and contribute."

Douglas also had misgivings about his own situation under an institute administration of Eastwoodhill – "I'm not wanted here

White birch, Betula pubescens.

White birch and Fly agaric toadstools, Amanita muscaria, *a common feature in autumn.*

Lawson cypress Chamaecyparis lawsoniana *beside The Cathedral, Cabin Park.*

Orange bark myrtle, Myrtus luma *or* Luma apiculata,
on the Redwood Path, Homestead Garden.

A seedling Burmese cherry **Prunus cerasoides** *along the Burma Road, Cabin Park.*

The autumn-flowering Burmese cherry in April 1995 below The Highway, Douglas Park.

Douglas-fir Pseudotsuga menziesii *near Windy Gap, Corner Park.*

Cabbage Tree Avenue, Pear Park, 1994.

Early spring flowering seedling hybrid of the bell-flowered cherry Prunus campanulata *between the trunks of manna gum* Eucalyptus viminalis *in the Gum Basin, Corner Park.*

if they take over; in fact I can't legally stay here. No one pays me anything and how the hell am I to live? Create Eastwoodhill and then be tossed out in the gutter to die." He ruled out, as an alternative, any possibility of selling to a farmer: "If I sell to a farmer he'll cut every tree down to grow grass. They're all on poor pumice land and all good land is farm."

Twice he mentioned the possibility of approaching Sir Henry Kelliher, Auckland-based founder of Dominion Breweries, who was noted for his patronage of the arts, but there is no record of his ever writing to him. He did write to then Prime Minister, the Rt Hon Keith Holyoake, inviting him to visit Eastwoodhill, but the PM politely declined because of other commitments. Douglas said he would like a man like Jack Goodwin of New Plymouth to run the property "but I think he feels Taranaki is his home". There was mention of Charles Puddle, head gardener for Lord Aberconway at Bodnant, North Wales – "He's about as good a plantsman as there is in Britain and ready to come if the salary is adequate but I can't suggest anything till the institute says they'll accept my offer."

But he had been developing a remarkable plan in the form of an Australian, Basil Keir, who had acted as his chauffeur during his travels in Britain and on the Continent in 1954. Douglas told Graham: "I've just had a lad staying with me who drove my car (and me) all over Britain and Europe. He was the perfect companion and driver and knew exactly when to speak and when to be silent." Basil had been a member of Pukeiti for 10 years – clearly signed up by Douglas during their travels. Douglas added: "He came over from Sydney to see me. He'd never been in New Zealand before and may become a New Zealander."

Douglas wrote again that "the young Australian" had offered to buy Eastwoodhill and, most important, to look after him – "He was like a son to me on our European journey and I'd be in good hands. I've no one near me interested and I'm mighty lonely at times – always." Basil was to pay a "substantial deposit", give Douglas £1000 a year, three meals a day, and look after him in his old age. He would keep Bill Crooks on and acquire more farm land as the present acreage was not economic.

Meanwhile, perhaps not coincidentally, Douglas gave the institute a deadline of February 28, 1965 to accept his offer – "After that date it will cost them £60,000."

Basil Keir arrived back at Eastwoodhill on February 4. "He has come over to see me and my lawyer and sign up. I'm afraid he's in a bit of a hurry as I can't do it till March 1. However, that he's flown over shows he's in earnest." Douglas observed, however, that when he travelled with him for five months he was a 20-year-old bachelor; now he was 30, married with three daughters – "That's the only part I'm afraid of. I don't know his wife and kids." Oddly, considering the nature of the proposed arrangement, he added that his friend seemed to want to live part-time in New Zealand and part-time in Australia.

It is clear that the plan was a pipedream, born out of Douglas's desperation and loneliness. He wrote to Graham on February 23 that during the Australian's visit "I got him to state his offer to the lawyer. Afterwards we both thought he wanted it given to him as I was to get £5000 for lock, stock and barrel. House, contents, all." The day after that letter, Basil Keir was still on his mind: "He's not really a gardener. It's a noble offer he's made and I don't think he could afford more. Should I think of Eastwoodhill or myself in my old age?" The final reference to the Australian in his correspondence with Graham came on March 24: "I don't think I'll sell to my young friend but I will try now to sell to Bill Crooks

This 1960 aerial view shows the newly-planted Orchard Hill (left). The garage, nursery and work sheds (right of centre) are now the site of the Douglas Cook Centre. On the front of a mounted copy of this photo Douglas had written, 'the heart of Eastwoodhill, the beauty lies beyond'. On the reverse was a quotation by Parkman used by his mother's family, 'He who would do some great thing in this short life must apply himself to work with such concentration of his forces as, to idle spectators, who live only to amuse themselves, looks like insanity'. (Photo Brian Sherriff, Wade Studios)

for whatever he can scratch together over £5000. Bill (though on wages) has built up Eastwoodhill just as much as I have and I know loves it and thinks things out and suggests them to me."

Ruefully, he observed that the "poorest farm around here" had recently changed hands at £37 an acre – "Poor and no water. Here we have good farm land, a permanent stream and 12 considerable never-dry dams. I started here 31 March 1910 and almost all this place earned me has gone into it. Trees, roads, paths, dams, buildings, water system – not to speak of a quite considerable library."

Douglas developed further the idea of selling to Bill Crooks by including in the equation Rob Bayly, the Gisborne nurseryman who had suffered heavy financial loss through Douglas's nursery scheme for Eastwoodhill in 1961. He wrote on June 15: "I had a lawyer out here with a friend who knows my wishes and we talked over my giving the place to Bill and Rob Bayly jointly. He and Bill can run a nursery and the farm and preserve the Park here. I've offered them practically all I've got here as a gift jointly to them both. Neither has any money but Bill can sell the sections I gave him at Wainui. You see, each will have to pay £1050 or £1100 gift duty as the place is valued at £12,000 - £14,000. Then, the snag is I've got to live three years in order to save some other tax of about £600/£800. So, as I don't expect to live three years I'll have to provide that too to be sure they can stay here.

Jo & Bill Crooks shortly after their departure from Eastwoodhill in 1974 (Photo Janette Crooks)

"Bill would be hopeless here alone. He's not a manager and would never make a boss but he's a good worker. Bayly is both and is a mechanically-minded man. Can look after all machinery. They'll do alright here and they get on well together and I get on well with both. I'm the only snag. I still want to live here but who is to look after me? I can't go on looking after this house. It's the women I don't get on with. Neither has meals at fixed hours and I like regular meals. It looks as if I may have to still get my own."

He added that he must keep The Jungle at Wainui in case he needed to be near a doctor.

Meanwhile, as Douglas had continued to develop his own schemes, and pleaded with various friends to take over Eastwoodhill, Colonel Gambrill maintained contact with the institute despite the fact that his client's "ultimatum" had long passed. As part of the institute's assessment of the value of the arboretum botanically, it wrote to Douglas with a list of questions concerning his plantings. His long reply included a typically terse comment: "You ask me to mention rare trees here. The place is full of them." But he went on to say that "the Forestry Department in Rotorua have a fairly good list of what is here and I think Mr W.

Sykes of Christchurch appreciates a lot of what he saw here and nowhere else in New Zealand. We still have our shade house and quarantine area full of rare imports not procurable in New Zealand."

Then he added: "I am certain that, put to the public of New Zealand, generous funds can be and will be raised."

In one letter, expressing his bitterness towards the institute, he wrote: "I'm now halted in my stride as it were, one foot in the air and don't know whether to go forwards or backwards. I've had a tough life to live in many ways but in all my adversities I've steadily if slowly gone Forward. That one word is the family motto on my Mother's side and the Miller family certainly went forward. Early in life I more or less made Kipling's 'If' my plan for living. How often I've had to think hard and keep on repeating its lines."

Through all of the frustration and the sparring with the institute there had been one significant glimmer of hope: Colonel Gambrill reported to his client that if the institute were successful in raising funds for a trust, a Gisborne man was prepared to make a substantial contribution. That man was H.B. (Bill) Williams, businessman, farmer and Angus stud breeder.

Months passed. No positive proposal had arrived from the institute. Then, on June 29, 1965 Colonel Gambrill rang Douglas with the news that he had a proposition to put to him. Douglas, apparently not understanding the importance of the call, replied that it was too late – he had made "final arrangements" to hand over Eastwoodhill to Bill Crooks and Rob Bayly. The Colonel insisted on a meeting. And Douglas learned that Bill Williams had offered to buy Eastwoodhill for £24,500, including the land, buildings, livestock, plant and vehicles. Bill Crooks was to stay on permanently, with a place for Rob Bayly if he wished; Douglas would stay in the house for life, and Bill Williams would organise the running of the farm.

Douglas accepted and, with the prospect of a large sum of money falling into his lap, began in his mind to spend £2000 on a new car, to reblock his house and Bill's, to put in an "inside lav" for Bill and to paint both houses.

He told Graham Weston: "I can't tell you the relief this is to me. I've decided to burn sox with no soles and get a pair of new slippers. Colonel Gambrill says go for a good trip on the Oriana. Could do that with pleasure too. "

8 A Soldier's Farewell

Douglas Cook had a heart attack in mid-1965 and it is clear that he never fully recovered from it. His letters to Graham Weston from that time form a saga of frustration and of frequent admissions to Cook Hospital as well as periods of recuperation in the Morris Convalescent Home. Of the attack, he wrote that his friend Brian Sherriff raced out from town to his aid having previously rung a doctor to arrange his admission to hospital – "We packed things and were at the hospital in little over half an hour."

Brian, a keen amateur photographer, had met Douglas through his hobby though he was to have another role in the Eastwoodhill story as an auctioneer with an interest in antiques. By coincidence, he was the son-in-law of Fred Barwick, proprietor of Barwick's Auction Mart, who was president of the Wellington Mounted Rifles Association and had met Douglas many times at association reunions. Douglas had also known Brian's wife Joyce for many years as she had been a staff member of Graham and

Pear Park in the late 1950s shows the part of Eastwoodhill as shown in the photograph in Chapter 1. Bill Crooks cottage is in the centre, with the circular Crooks' Walk to the centre left. The Douglas Park Nursery, a commercial venture of Douglas Cook's in the late 1940s to early 1950s is on the extreme left. The woolshed was built by Emmersons in 1928 for £240. Prior to this neighbours' sheds were hired. The radiata shelter belts featured here were cleared by H.B. Williams after 1965 to provide the open spaces of today. (Photo Brian Sherriff, Wade Studios)

Dobson, the accountancy firm which handled Douglas's bookkeeping. Douglas would arrive at the office with a bunch of flowers for her and another for Dorrie Hawkins who handled his account.

Brian combined his love of photography with a passion for flying. Though he never had a pilot's licence himself he went frequently on flights with a Dutch friend, Hank de Heus, a topdressing pilot. Brian took aerial photographs of many country homes and on one occasion in the early 1960s made a special run over Eastwoodhill. Later he rang Douglas, who then made a trip into town to see the prints at the auction rooms. Brian recalled: "When he viewed the finished photos he was so thrilled with them he told Mr Barwick that he would like me to go out to Eastwoodhill and look at Persian rugs and silver. He wanted me to sell some on his behalf so that he could raise money to pay for the many photos he particularly wanted to purchase."

From that time Brian assisted Douglas in selling antiques at current values. He respected Douglas's own expertise: "He was a man of considerable taste who must have learnt from the vast collection of books in his library. He was a very worldly person, not only interested in plants and trees but any item of beauty and works of art. I regarded Mr Cook as a true friend – a very kind-hearted and genial man with a good sense of humour. He enjoyed a joke but being the perfect gentleman would not have appreciated anything in the least smutty in the presence of a woman."

Brian Sherriff became a frequent visitor at weekends. On occasions Douglas would tell him to bring Joyce and "the boys" (Garth and Brett) to afternoon tea. On the first occasion it was

A photograph of a frail Douglas Cook in spring 1965 as he sits on the steps of Palm Terrace and chats with a visitor. He holds a branch of the violet-scented Betchel's crab, **Malus ioensis plena** *which he had planted in Douglas Park.*

rather formal: "Mrs Sheriff, would you kindly pour the tea?" She soon became "Joyce" but the best silver tea service and chinaware were always used. Once, on Douglas's birthday, they paid him a surprise visit, taking along Dorrie Hawkins and another acquaintance, Lil Godfrey – "He was so delighted to see old friends who cared about his special day that he was quite tearful." On that occasion they had stronger drink than tea – two drinks in fact. But Douglas would not allow Brian a second because he was driving.

The work-horse at Eastwoodhill for many years was an old 14hp Morris truck purchased from Allen Bros & Johnstone for £125 in 1942. It had been previously owned by D.J. Barry's Brewery. In the place of a tractor and trailer, this old faithful performed many tasks about the garden. It was also used to take flowers and foliage into Gisborne to the Army Hall for the Horticultural Society's flower shows for decoration of the hall. The truck is shown here with Bill Crooks and his son Robin at Rock Point.

As the friendship between the two men developed, Brian took Douglas on a number of trips, locally to see gardens and sometimes out of the district. There was one journey to Rotorua to contact prospective buyers for his antiques, and a pleasure trip around the East Coast with an overnight stop at the Te Kaha Hotel.

From The Jungle at Wainui, Douglas gave the Sherriffs a special creeper, a white lapageria, which he knew would not thrive at Eastwoodhill. It flourished at the Sherriff home in town. As his health began to fail, Douglas wanted to give Eastwoodhill to Brian and Joyce. However, it would have been impossible for the couple to have managed the property or paid for its maintenance. But there is evidence that Brian added his own voice to the plea for Bill Williams to buy it.

After Douglas had a second heart attack, arrangements were under way for the Crooks to move into his home so that they could be constantly on hand. A new hot water cylinder and a new stove were being installed, which Douglas saw as an unnecessary expense for Bill Williams. He could not resist a small grizzle to

Graham: "Bill (Crooks) cleaned up the frightful mess four men made yesterday." And on December 30 he said he was trying to confine himself to five rooms after being used to 11 – "Selling such a lot, but I'll have plenty left."

Despite the state of his health, and because of it, the future of Eastwoodhill was uppermost in his mind and he was concerned that updating of his plant catalogue was not progressing: "I can't touch my catalogue. Brain won't work and I've been seriously thinking of asking you (Graham Weston) to take my two copies and make the final copy, correctly alphabetical, leaving four to five blank lines between each for additions. This would be a paid job. Would you?" In a letter to Bill Williams, Graham mentioned the catalogue but explained that it was a big task. To do it properly would mean physically checking every plant in the garden – a fulltime job for a fairly long period which he could not undertake. However, he said that he might make a visit to Eastwoodhill in the summer with Bill Sykes and they could discuss it with Douglas.

During a forestry institute visit, Douglas accompanied the group by car on a tour of the arboretum. The fact that he could no longer stroll among his trees and gardens depressed him – "I'm not allowed to walk hills or grades. I'm properly hampered. The garden I can walk quietly in and there is beauty. I love it."

His letters, as always, contained frequent references to plants:

"Regarding sick spruces, I think perhaps we try to grow too many from high altitudes and too many from cold climates while *Cedrus atlantica* from Morocco and *Abies pinsapo* from Spain appreciate our warm, dry climate. Even *Picea omorika* does splendidly. I haven't nearly enough of them. We just can't grow happily here everything from hot and cold climates and high mountains to sea coast. We try, and fail, and try again."

And another: "I think my failure with conifers is largely that they hail from hot dry and cold wet lands and I can't please all. In fact, only a few. Someone after I'm gone can have another go at them."

On November 21, 1966: "Spring was marvellous. Never have I seen such profusion of bloom, not just in magnolias but in every tree and shrub. Some still lovely. Deciduous azaleas magnificent. Daffodils in sheets of colour. That was a wonderful send-off to a working man's life."

"So many plants from England not planted out. Hundreds of pounds worth. Bill got what he could done this year. The word is that Hillier has sent by air 20 betula and a rhododendron I ordered and forgot to cancel."

Two conifers that perform exceptionally well under Eastwoodhill conditions are tall **Cedrus atlantica** *'Glauca', the blue Mount Atlas cedar on the left, and the narrow spire of* **Picea omorika,** *the Serbian spruce, growing together at the lower end of the Daffodil Patch. (Photo Garry Clapperton)*

Douglas Cook's health failed him one night. Choking for breath, he had knocked on the bedroom wall and Bill stood by until the following morning when Douglas was admitted to Chelsea Private Hospital. Douglas spent his birthday there on October 28, 1966. Bill and Jo Crooks paid him almost constant attention. Douglas later wrote that his doctor wanted him to engage someone to look after him "but I said NO. Bill will look after me. HE and his wife do all I need… Bill is like a most attentive son to me."

June Murphy took him for a drive around the arboretum in her Morris Minor shortly before he was admitted to hospital for the last time. Douglas took his walking-stick and pointed it to indicate where he wanted to go.

He died in hospital on April 27, 1967 in his 83rd year. In a notice in the Gisborne Herald, Fred Barwick invited ex-members of Wellington Mounted Rifles, 1st NZEF, to attend the funeral of Tpr W.D. Cook. Dress: Black and white tie. The service at Evans Chapel next morning was to be followed by private cremation at Hastings which had the crematorium closest to Gisborne at the time. The tie was the official emblem of the Mounted Rifles Association and was appropriate for the occasion. Douglas had worn his proudly at reunions. Brian Sherriff recalled that Douglas valued the fellowship of wartime comrades and members of the R.S.A. "but feared they found him different".

Roland Graham, who died in Gisborne in 1994 in his 104th year, quipped on his 103rd birthday: "I remember a big car-load of us mounted riflemen going to a funeral once and Fred Barwick asked who would be the last one to wear the black and white tie. Little did I know it would be me." Apart from reunions held at Eastwoodhill, Roland and his wife Dora sometimes called there at weekends. Roland once recalled: "My wife was fond of gardening. She never read a book that wasn't about gardening and when she and Douglas got together I scarcely could get a word in. They held the floor." Douglas would give Dora cuttings and plants for her garden. Roland remembered Douglas showing them a young tree which he believed to be one of only three existing outside of China – one at Kew, one in the United States and the third at Eastwoodhill.

Roland, who considered Douglas "an artistic rather wonderful fellow" was born in 1890 at Mangatoetoe Station, a property at Ngatapa farmed by his parents, not far from the land that was to become Eastwoodhill. Appropriately, his son Don became a member of the Eastwoodhill Trust Board in the 1980s.

On the day after Douglas Cook's death the Gisborne Herald

The Veitch Memorial Medal from the Royal Horticultural Society of the U.K. was awarded in recognition of Douglas Cook's efforts over 45 years for service to horticulture. Douglas had successfully nominated others in New Zealand for the same award including Jack Goodwin and Victor Davies

This medal was awarded to Douglas Cook by the Poverty Bay Horticultural Society in 1952 for his rhododendron and azalea display. Douglas provided the foliage and flowers on many occasions to decorate the stark interior of the Army Hall for flower shows, bringing material down from Eastwoodhill on his old Morris truck.

published an obituary which covered his gift of land at Pukeiti and the establishment of the Pukeiti Rhododendron Trust; his achievements at Eastwoodhill "known among horticulturists all over the world"; his gift of the Silver Challenge Rosebowl for competition among members of the Horticultural Society; and his fund-raising for various local organisations through his open days.

"A plantsman with the soul of a poet and the vision of a philosopher" – those were the words used by New Zealand Gardener magazine to describe William Douglas Cook in the foreword to an article by him in its issue of January 1, 1948. The passage deserves to be quoted in full: "The real gardener has a desire to turn the wilderness into a Garden of Eden, and to make the Earth a more beautiful place for his having lived on it, but to few is given the opportunity to carry such a dream into effect. Some miles to the north of Gisborne, Mr W. Douglas Cook, a plantsman with the soul of a poet and the vision of a philosopher,

Douglas's Cook's homestead as it appeared in the mid-1960s. The brick walls and and circular steps provide a formality to the immediate environs of the house not repeated in the remainder of the park. (Photo Brian Sherriff, Wade Studios)

had such a dream, and has subordinated all his other interests to the creation of a garden that promises to become a national asset."

Douglas had been made a Fellow of the Royal New Zealand Institute of Horticulture in 1948 and in 1966 was belatedly elected an Associate of Honour, the highest RNZIH award. In 1965 the Royal Horticultural Society, of which he was a member, honoured his services to horticulture with the award of the Veitch Memorial Medal. In an obituary in the Pukeiti newsletter, in May 1967, J.W. Goodwin wrote: "Douglas Cook played a noble part in furthering horticulture in New Zealand and it may justly be said that by his efforts he left this world just so much better. Surely his memorial will be found in the trees he planted."

Perhaps, in a sense, Douglas had written his own obituary when in October 1965 he mused: "I've had a busy life and have lived as good a Christian life as I could and have set my service to man high. I think I've accomplished something which will be more and more appreciated as time goes on. I've suffered all the setbacks any man could. Poor food, living alone, depression through lack of money and at times ill health through worry… The family motto being Forward, I've never let it down. If to be poor is to prove oneself a failure, then I've failed. But I've left a heritage behind me for future generations of New Zealanders to enjoy… The finest country and the finest people in the world."

Two 1954 scenes of Douglas Cook's 'new drive' that he built in 1936. On the left is a scene of the bridge near the present exit gate with the Californian redwood in the centre. It is today over 37 metres tall. On the right the scene is of a young blue Mount Atlas cedar adjacent to the Wet Lawn. (Photos Douglas Elliott)

Douglas Cook's Three Homes at Eastwoodhill.

The small two-room cottage completed by May 1910 was Douglas Cook's first home. Peach and Stone were paid £74/2/- and Cave Bros £14 for the building. Greaves did the cartage.

Douglas Cook's bedroom part of the cottage. Though the wall is adorned with ladies' photos, the two in the centre are probably family portraits. A teddy bear is on the table beneath the window.

In 1914 the hill was levelled to form the House Terrace and the first part of the new house built. £130/9/9 was spent. A lean-to had been built on to the cottage in 1912 and this was relocated to form the basis of a wash-house. When this was demolished in 1994 several copies of The Poverty Bay Herald from May 1912 were found lining the floor under the linoleum. The same portion was Bill Crooks' bedroom until he married

Two views of the north side of the new house, showing (above and right) how Douglas Cook added piecemeal out to the eaves and a box-like porch. He continued to enlarge the house in this manner.

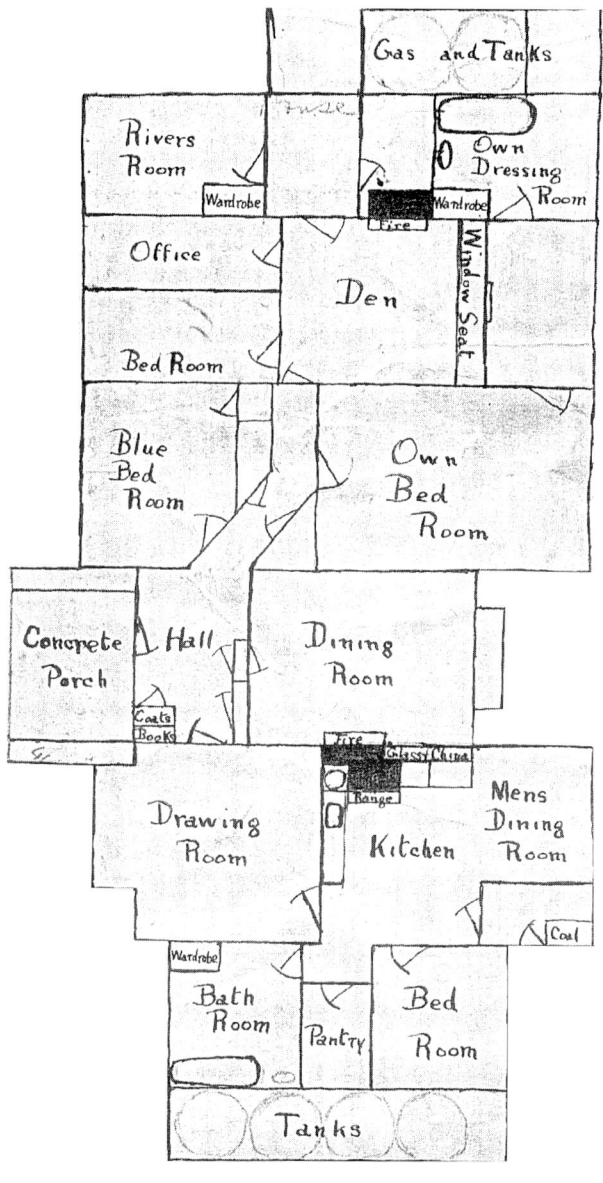

Plan of the 'warren' as drawn by Douglas Cook and included in the Homestead Garden layout plan in 1926.

Prior to the arrival of power on July 8 1929, house lighting was done with acetylene gas, the tower being at Douglas Cook's end of the house.

A photo of Gisborne hangs on the wall in the dining room as well as various family photos on the mantel-piece. The original photo was from a glass-plate negative so it is possible to clearly see the people and in another photo taken in the Den, Douglas Cook is clearly seen as a young boy. All this was lost in the 1931 fire.

The existing house, Cook's third dwelling was built in four stages, the first section being ready by June 1931. Shortly after he built his bedroom and office separate from the house, so perhaps strain was already being felt in the marriage.

The north side of the existing house.

The complete house in about 1937, the smaller window in the lounge was replaced with a larger pane later. The door from the library on the right has never had steps and even today doesn't have a real purpose.

The second dwelling prior to the fire in 1931 taken from the House Terrace.

The same view as top right in 1996.

9 The Hand of Fate

It is tempting to believe that fate stepped in to save Eastwoodhill – fate in the person of Colonel Reg Gambrill. Bill Williams had never met Douglas Cook before negotiations began for the purchase of the property and he had visited Eastwoodhill previously only once, as a member of the public on one of Douglas's Daffodil Days. Certainly, it was a fateful circumstance that in 1965 Colonel Gambrill happened to be acting as lawyer both for Douglas and for Bill Williams; otherwise Bill would not have been aware that the arboretum was in serious danger. There is also the fact that "the Colonel" (as everyone knew him) had links with the Williams family dating back several decades. When Douglas Cook was urgently seeking help to preserve his park in 1965 it was natural for Reg Gambrill to think of Bill Williams. He knew that Bill had the right attributes – business acumen, the financial resources to buy the property, farming expertise and,

Frimley was the Williams family homestead in Hastings, the home of James Nelson and Mary Williams. In about 1905 Douglas Cook bought 10 acres of J.N.'s Frimley peach orchard on Omahu Road. J.N.'s son H.B. was a director of Horton's Nurseries, from which Douglas Cook bought most of his early trees, and also father of H.B. (Bill) Williams who bought Eastwoodhill from Douglas in 1965. Another of J.N.'s sons was A.B. who was involved in the Waipiro Bay operations where Douglas worked for a time prior to 1910. Douglas knew the garden at Frimley and its 40-year-old trees may have been an inspiration to him. The Frimley homestead burned down in 1950 and the trees and grounds presented to the people of Hastings as a park in 1951.
This photo was taken in 1936. (Photo NZ Aerial Mapping)

above all, a generous spirit that had made him a major donor towards community needs in Gisborne. And, though not a botanist, he continued a family tradition of admiration and respect for trees and plants.

For Bill Williams, the proposal to buy Eastwoodhill was not a business proposition in the commercial sense. Not only did he face the challenge of working the farm profitably; he would have to find a way of covering the costs of maintaining and hopefully enhancing the park. During the negotiations he had read reports by Graham Weston, Bill Sykes and other experts. Convinced that Eastwoodhill was indeed a national treasure which must be saved, he was ready to sign – "I was delighted to be able to assist." Looking to the future, Bill Williams now had a tall dream of his own.

Meanwhile, Graham Weston had been busy convincing fellow plantsmen that Eastwoodhill should be preserved and had arranged with Douglas to bring a party from the Forest Research Institute on a group visit. In the institute newsletter circulated in June, 1965, he quoted comments by Bill Sykes after Bill had spent several days at Eastwoodhill in February and gathered 300 specimens for the DSIR Botany Division's herbarium.

"I should say that there is no place in the country with anything like the collection of north temperate woody plants," Bill Sykes had reported. "In fact I have not seen anything like it since leaving England. This collection is of immense potential value to me." Of the specimens he had gathered he added: "This only scratches the surface of what is at Eastwoodhill, I know. There must be few places in the world where so many of the rare coniferous genera are growing together."

Graham Weston added his own comments in the newsletter: "The big question at present is what is to become of Eastwoodhill. Mr Cook is over 80, and upkeep and development of the garden are now beyond him, but his one wish is to see his life's work carried on. Ironically, Mr Cook was himself the main driving force behind the development of the very successful Pukeiti Rhododendron Trust on the slopes of Mt Egmont and gave the land for it. Can something similar be done for Eastwoodhill? By whom? With what resources?"

Graham subsequently wrote to Bill Williams about the proposed visit by the institute group. Bill replied on October 7, 1965: "My intention in buying Eastwoodhill was solely to preserve it for the future and for the enjoyment of all interested in horticulture. Any visits of local and outside interested people are welcome." To another question raised by Graham he replied: "With reference to making any per capita charge to visitors, this would

Willow oak, Quercus phellos *in Douglas Park.*

Self-sown mamaku tree ferns Cyathea medullaris *grow on the bank near Rest ani be Thankful, Cabin Park. Fallen leaves of scarlet oak and common beech litter the path.*

Nurserywood, Pear Park, was lined with rows of red oak, Quercus rubra *in 1948 when it was the commercial venture Douglas Park Nursery.*

Beech leaves on the ground in the Black Forest, Cabin Park.

Scarlet oak and Japanese maple feature at Circus Corner, Cabin Park.

The H.B. Williams Walk along The Vista, Douglas Park.

Swamp cypress Taxodium distichum *and Blue Mt Atlas cedar* Cedrus atlantica 'Glauca' *at Rock Point Pond.*

Swamp cypress at Hillier's Pond, The Circus.

In Douglas Park 1962 looking over Blackwater and a young Aesculus indica *across to Lookout Hill on Rock Ridge.*

The Miniature Conifers in Douglas Park, now on the H. B. Williams Walk.

xix

The Beechwood, Cabin Park. A valley of common beech, Fagus sylvatica. *Planted 1936.*

Maidenhair tree, Ginkgo biloba *along Apple Way, Douglas Park. Planted 1963.*

Shag-bark hickory Carya ovata *on Linden Green, Pear Park. Planted 1963.*

Japanese maple, Acer palmatum 'Osakazuki', *beside Birch Hill Pond, Douglas Park.*

Garry Clapperton

of course be voluntary and most welcome. As you know, Eastwoodhill is not self-supporting by a long shot and any financial assistance towards upkeep and preservation would be most acceptable. I hope to go into ways and means of building up some fund which eventually will be big enough to bridge the gap between income (ex-farm) and costs."

The story of the Williams family in New Zealand began with the arrival from Britain of two missionary brothers, Henry and William. The younger brother William, great-grandfather of Bill, arrived at the Bay of Islands with his wife in 1826. He became known as The Educator. As well as teaching children in a general way, he also taught the Maori language, translated the Bible into Maori and produced in l844 A Dictionary of the New Zealand Language. He set up the first mission station at Waerenga-a-Hika in Gisborne in 1840 and became, in 1859, the first Bishop of Waiapu, the Anglican diocese embracing Hawke's Bay, the Gisborne district and the East Coast.

William's young son, James Nelson Williams, was brought up in Gisborne but began his farming activities in Hawke's Bay as a 20-year-old, later building his first home at Frimley, near Hastings. At the age of 46 he was to become a major figure in opening up the East Coast to farming. In search of land for settlement, he set out from Gisborne on horseback in 1883, riding 70 miles through rugged country to reach Waipiro Bay. He acquired sufficient leases to aggregate 39,000 acres and created Waipiro Station. The Bay became transformed into a busy though modest seaport. For the local Maori the activity provided gainful employment – bush-felling and post-splitting and later, shearing.

Coincidences and links abound in the Eastwoodhill story: It was, of course, J.N. Williams from whom the young Douglas Cook acquired a block of peach orchard and for whom he worked as a farm cadet in Waipiro Bay. J. N. Williams had two sons, Heathcote Beetham and Arnold Beetham, who carried on developing farm properties on the Coast, and established numerous business interests. They became known throughout the region as "H.B." and "A.B." When H.B. had a second son he had him christened by the same name as his own. But the youngster was always known as "Bill".

Bill was "brought up with trees". A.B. had built a home at Puketiti Station and planted extensively – eucalypts, Oregon pine and natives. Bill, who visited the property frequently from his childhood, was to observe: "He created a memorial to himself at Puketiti. As a family, planting was always an important part of establishing homes." H.B. also planted extensively; J.N. Williams had created Frimley Park, and a cousin of Bill's, Meyric Williams (who knew both Eastwoodhill and Douglas Cook well), was a tree consultant in Hawke's Bay. Bill once did some public planting of his own: When he felt that the district council had planted the wrong type of trees along the roadside outside his Turihaua stud farm he replaced them himself with pohutukawa seedlings.

It is doubtful if Bill Williams ever faced a more challenging task than Eastwoodhill. However, with his wife Elizabeth and children he enjoyed picnic outings to the property. With a cut lunch, slashers and saws they would drive out from Turihaua for the day. Wattle, broom and manuka abounded and everybody joined in, cutting scrub and burning it. That activity was only cosmetic treatment considering the magnitude of the task. But Bill had a plan: he believed that with expert supervision of the farm, and with good fencing to safeguard the tree garden, he could double the carrying capacity to about 1000 ewes. Increased profitability would create the funds needed to tidy the arboretum and maintain it into the future. He appointed Tim Lewis as a farm consultant. The scheme also relieved Bill Crooks from farming work so that he could spend more time in the arboretum.

The plan, however, was based on 1965/66 prices for lambs and wool. Those prices began to fall drastically while inflation was forcing up costs. There would have to be another way.

10 A Remarkable Document

Among the archives of Eastwoodhill is a remarkable document – 39 pages of single-spaced typing which, to the untrained eye, look like meaningless hieroglyphics. Much of the typing is in fact a form of code; numbers and letters represent positions on a grid map of the arboretum; other letters are abbreviations for types of plants, nurseries or organisations. The work is simply titled: Eastwoodhill Arboretum: Catalogue of Trees, Shrubs and Climbers. Within those pages is contained detailed information of almost 3000 different species and sub-species of trees and other plants.

The catalogue is remarkable not only for the scope and detail of the information it contains but also for the devotion of its author. Bob Berry, a tall, lean farmer from Tiniroto, on the inland route between Gisborne and Wairoa, is a plantsman with an international reputation, particularly as an authority on the oaks of Mexico. It was his sheer love of plants and his concern for Eastwoodhill that, in the early 1970s, led him to the conclusion that the specimens growing in the arboretum, many of them rare, must be listed. Without such a list, the true value of Eastwoodhill could never be known.

Bob grew up in Tiniroto on Hackfalls Station, named by his grandfather after a place he knew in Yorkshire. Bob's father was a practical farmer who only planted trees that would be of some use. As soon as his father died, Bob was able to develop a deeper interest in trees. He straight away began planting varieties for their beauty and botanical interest. A serious involvement with oaks began in the 1960s when he started to collect European, North African and North American species. Later his interest focused on the oaks of Mexico which is home to up to 150 different species – more than half of the oak species throughout the world. Through seed-collecting trips to that country he introduced about 35 Mexican oak varieties to New Zealand.

Through the years he had been building a substantial library of books on dendrology and became a member of the Horticultural Society. It was as a member of that organisation, taking part in a field trip in 1953, that he first visited Eastwoodhill Arboretum. Fascinated by the collection, he became a regular caller.

He discovered Douglas Cook's puckish sense of humour early in their association when he arrived to buy some cuttings of new American hybrid poplars, Schreiner hybrids which Douglas had advertised for sale. Bob chose 10 cuttings of different varieties but was taken aback when Douglas announced: "That will be five pounds." That seemed rather a steep figure to Bob. Then, as he handed over the note, Douglas added with a hint of a smile, "And now I'll sign you up as a member of Pukeiti." The five pounds was the subscription fee; the cuttings were free. Bob remained a Pukeiti member, happily paying his membership fee every year.

There was also a touch of humour, or rather irony, in Douglas's admiration of the Eastwoodhill pheasants. He told Bob he liked to see them "strutting around among the trees" and made it clear to local sportsmen that they were not to be shot. However, he had planted some valuable bulbs along each side of a path. The pheasants discovered them, scratched them out and ate them. Douglas quickly spread the word that it was now "open season" on the pheasants. But he never shot one himself.

The two plantsmen (who always remained "Mr Cook" and "Mr Berry") frequently discussed Hackfalls. Douglas would offer advice and make suggestions, such as planting distances. Bob took his advice, increasing the distance between his plantings – "I learned a lot from Eastwoodhill as I developed my own arboretum". Oddly, Douglas never paid a single visit to Hackfalls, but Bill Crooks used to call there in later years.

Bob had carefully labelled all his plants over the years and kept precise records. Following Douglas Cook's death he became increasingly concerned that labels from some plants at Eastwoodhill had been lost and others transposed. Douglas had rued an experience with mischievous youngsters. He had allowed Boy Scout groups to roam in the arboretum – no doubt his way of encouraging youthful interest in trees. When he discovered that boys had removed labels from some plants and attached them to others the Boy Scout visits ended abruptly. Douglas also kept records, but after his death many were found to be inexact and lacking precise detail as to location. No labels carried planting dates; they could be established only through nursery records. It had been his intention to create a list, but planting itself became his priority as his health began to fail.

When, in 1971, Bob prepared to begin his mammoth task of listing the plants of Eastwoodhill, he knew that some system was necessary to establish the location of plants. Colin Pilbrow, an architect with the Gisborne firm Glengarry, Corson and Pilbrow, prepared at no cost a map from aerial photography and overlaid it with a grid system. Bob began a routine. On at least one day a week, depending on weather, he would set out with a cut lunch and a notebook in his knapsack and drive 60-odd kilometres to Eastwoodhill. Methodically, he began within the limited area equivalent to one square on the grid. Bill Crooks was an invaluable aid, having assisted with most of the planting over a period of 40 years. In earlier years Bill had difficulty coping with the Latin names of plants. But with the acute foresight of a man with a vision of Eastwoodhill's future, Douglas had begun in the later years of his life to coach Bill in correct botanical terms and their pronunciation. Overgrowth did not help Bob in his work but he was to recall: "Bill knew where things were. He would scrabble amongst the fern and uncover a plant that could have been missed."

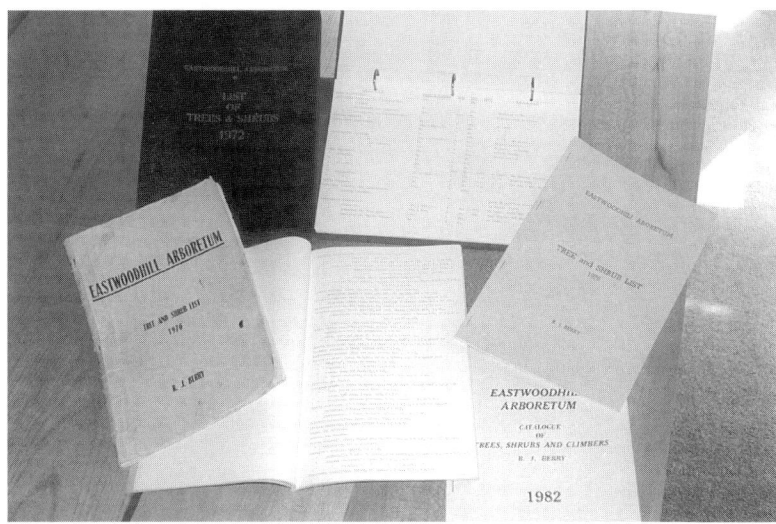

Until 1986, Bob Berry was solely responsible for the catalogue of the Eastwoodhill collection. He issued various updates on his original 1972 version which had been issued in a foolscap form in a ring binder folder (at rear of photo). The heights and diameters of important trees were included.

It was a long, painstaking process which extended over more than a year. If Bob was unable to recognise a plant he would take samples of leaves, berries or flowers back to Hackfalls and identify them through the volumes in his library. Sometimes that system meant returning to the same plant at a different time of the year when it was flowering or fruiting. As sections of the arboretum were tidied, more material would appear – "odd plants kept

S P E C I E S	HEIGHT X DIAMETER	DATE	SITE	FRUIT	R E M A R K S
Quercus X ludoviciana.	18m. x 68½cm.(T).	1949.	D12.	Yes.	A fine specimen. The most vigorous, oak at Eastwoodhill.
" lusitanica?.	1m.	1955.	J4.		Appears to be dead above graft, and reverted to Q.ilex stock.
" lyrata.		1970.	H11.		In garden. See also Q.stellata.
" macrocarpa.	9m. x 15cm.(T).	?	I4.	Yes.	No label. Poor form.
" "	5m. x 11½cm.(T).	1951?.	S5.		The best shaped specimen.
" "	5½m. x 7¼cm.(T).	1948.	C11.		A slow growing specimen.
" marilandica.	8m. x 21¼cm.(T).	1949.	B12.		Leaves usually turn red in Autumn.
" mirbeckii. SeeQ.canariensis.					
" mongolica.		1970.	H11.		In garden. Seedlings may be hybrids.
" myrsinifolia. See Q.variabilis.					Tree under this name appears to be Q.variabilis.
" nigra.	13m. x 50cm.(T).	1948?.	P7.		Fine tree. No acorns seen yet.
" " ?.	14m. x 51¼cm.(T).	?	I4.		" " No label.
" palustrus.	25m. x 61cm.(T).	?	E4.	Yes.	None of this species have labels.
" "	20m. x 59cm.(T).	?	G10.	Yes.	Beside lawn. Many others. Fine trees.
" " cv. 'Horizontalis'?.	10m. x 48cm.(T).	1955?.	Q9.	Yes?.	No label. 12m. spread of branches.
" phellos?.	12m.	1947.	O7.	Yes.	Divided trunk. Q.leana on label.
" petraea cv. 'Mespilifolia'.	17m. x 46cm.(T).	?	J3.		Fine tree.
" " " "	14m. x 30cm.(T).	1955?.	P8.		" "
" petraea cv. 'Purpurea'.	3m.	1955?.	I8.		Not very vigorous.
" phillyraeoides.	2½m.	1955.	P3.		
" pontica.	1¼m.	1955.	Q5.		
" ? .	12m. x 35cm.(T).	1948?.	D10.	Yes.	Appears to be a hybrid between Q.robur and Q.prinus?.
" X pseudo-turneri.	3½m.	1955.	Q8.		Spreading, bushy tree.
" pubescens. (lanuginosa).	8m. x 20cm.(T).	1948?.	D12.	Yes.	Poor form.

P.(57)

A sample page from the 1976 catalogue.

turning up". He also included measurements in his list – estimates of heights and spread in the case of tall trees, actual measurements for low branching trees and shrubs.

The field work extending over all those months into 1972 was not the end of Bob's labours. He now had to collate his notes into a detailed list. He travelled to Rotorua to discuss methods of presenting the list with Graham Weston of the Forest Research Institute. With the substantial amount of detail that needed to be incorporated into each line, handwriting was out of the question. Bob, whose fingers had never touched a keyboard, bought himself a portable typewriter for the task. Graham Weston was enthusiastic about the project in determining the true value of the Eastwoodhill collection. "Otherwise," he wrote in a letter to a friend, "it is a bit like continuing to pride oneself on owning a magnificent stamp collection, locked away but allegedly containing some known rarities, and attempting to assess its value without checking..."

The value of the list became apparent immediately when Bob sent copies to interested plantsmen, botanists and organisations throughout New Zealand and abroad. One of the recipients was the Arnold Arboretum of Harvard University. The director, Richard A. Howard, replied on July 13, 1972 with this observation: "The collection described is an interesting one and indicates an amazing variety of hardy types. If you or anyone else prepares herbarium specimens to document this collection may I indicate our interest in having duplicates of such material. We maintain an herbarium of plants under cultivation with worldwide representation, but with relatively few from New Zealand. Any specimens from Eastwoodhill or comparable collections would be of interest to us." Thus was forged another link in the international dendrology chain.

Of greater immediate interest was a response from Sir Harold Hillier, senior member of Hillier and Sons of Winchester, England, who had been the main suppliers of plant material to Douglas Cook. His reply, on July 25, 1972 indicated that he had lost the stock of a number of important species and would "very much welcome seeds from Eastwoodhill." Bob Berry was excited. He wrote to an enthusiast, Bill Way, pointing out that the two letters "indicate an interest in the collection which could be interpreted as an indication of its uniqueness." He continued: "The interest shown in particular by Mr Hillier in certain plants could mean that these may no longer be growing in Britain, otherwise he

A 1963 aerial photograph showing Douglas Park. The bare slope of Rock Ridge is prominent above the Rock Point Pond (centre). Birch Hill Pond (left foreground) has only recently been completed and the water is still milky from the settling ash. The bare, upper left side of Rock Ridge is the Hillside Pinetum of today, where over 100 species of pines are planted.

A well-known feature of Eastwoodhill are the Patula pines Pinus patula *on Ocote Flat, Cabin Park. The patterns are created when winds cause breakage of small brittle branches.*

Foliage of Pinus patula

The way to admire the canopy patterns of the Patula pines at Ocote Flat (see opposite).

The multi-stemmed Bhutan pine, Pinus wallichiana *of 1928 on Dry Ridge, Corner Park.*

A discussion under the Patula pines on Ocote Flat, Cabin Park.

Children, autumn and leaves, what else but…

List of Genera:

The list of genera at Eastwoodhill is a good indication of the importance of the arboretum as a collection not only in New Zealand but in the world.   There are 168 families in the Eastwoodhill collection as of December 1996.

A **family** such as Fagaceae contains seven genera, Castanopsis, Castanea, Fagus, Quercus, Lithocarpus, Cyclobalanopsis, Nothofagus, Chrysolepis .   These seven genera have about 1050 **species** between them.

Cultivars or varieties are individiual plants of a species which are selected and named for some special feature, such as form of tree, leaf shape, flower colour etc.

Sometimes species cross to give a hybrid. There are also a few crosses between genera, but this is quite rare.

An example of a botanical name for a tree at Eastwoodhill, such as *Quercus robur* 'Filicifolia'

Family	Fagaceae – being the beech family
Genus	*Quercus* – being the oak genus
Species	*robur* – being the Common oak of Europe
Cultivar	'Filicifolia' – being the cut leaf form of the Common oak

Following are the genera in the collection at Eastwoodhill representing over 3200 different species, cultivars, hybrids and varieties (the numbers are marked after each genus.).

Genus	No.	Genus	No.	Genus	No.	Genus	No.
Abelia	4	Broussonetia	1	Cladastris	2	Dodonaea	2
Abeliophyllum	1	Brugmansia	2	Clematis	22	Dombeya	2
Aberia	1	Buddleja	31	Clerodendrum	3	Doryanthes	1
Abies	40	Bupleurum	1	Clethra	7	Doxantha	1
Abutilon	11	Buxus	3	Clianthus	1	Dregia	1
Acacia	15			Colletia	2	Drimys	2
Acanthopanax	1	Calliandra	1	Colquhounia	2	Duranta	2
Acca	1	Callicarpa	10	Coprosma	8	Dysoxylum	2
Acer	162	Callicoma	1	Cordyline	6		
Acokanthera	1	Callistemon	7	Coriaria	1	Echium	2
Acradenia	1	Callitris	8	Cornus	26	Edgeworthia	2
Actinostrobus	1	Calluna	1	Cornyocarpus	1	Ehretia	2
Adenandra	1	Calocedrus	2	Corokia	3	Elaeagnus	9
Aesculus	19	Calodendrum	1	Corylopsis	7	Elaeocarpus	2
Agathis	1	Calycanthus	2	Corylus	9	Embothrium	2
Agyrocytisus	1	Camellia	240	Corynabutilon	1	Emmenopterys	1
Ailanthus	2	Campanula	1	Cotinus	3	Enkianthus	4
Akebia	1	Campsis	2	Cotoneaster	14	Erica	3
Alangium	2	Camptotheca	1	Craibiodendron	1	Eriobotrya	2
Alberta	1	Campylotrophis	1	+ Crataegomespilus	2	Eriogonum	1
Albizia	2	Cardiocrinum	1	Crataegus	24	Eriostemon	1
Alectryon	2	Carmichaelia	2	X Crataemespilus	1	Escallonia	6
Aleurites	1	Carpenteria	1	Crinodendron	1	Eucalpytus	9
Alniphyllum	1	Carpinus	11	Crotolaria	1	Eucommia	1
Alnus	46	Carpodetus	1	Cryptocarya	1	Eucryphia	6
Aloe	2	Carrierea	1	Cryptomeria	10	Euonymus	19
Aloysia	1	Carya	7	Cudrania	1	Euptelea	2
Alphitonia	1	Cassia	1	Cunninghamia	3	Eurya	4
Alyogyne	1	Cassinia	1	Cunonia	1	Euscaphis	1
Alyxia	1	Castanea	2	Cuphea	1	Evodia	4
Amelanchier	7	Castanopsis	2	Cupressus	34	Exbucklandia	1
Amorpha	1	Castanospermum	1	Cyathea	2	Exochorda	5
Ampelopsis	5	Casuarina	1	Cyathodes	2		
Angophora	1	Catalpa	10	Cydonia	1	Fagus	15
Aphananthe	1	Catha	1	Cynabutilon	1	Fatsia	1
Aralia	5	Ceanothus	3	Cyphomandra	1	Ficus	1
Araucaria	4	Cedrela	2	Cyrilla	1	Firmiana	1
Arbutus	4	Cedrus	13	Cytisus	1	Fokienia	1
Ardisia	1	Celastrus	1			Fontanesia	1
Aristolochia	2	Celtis	16	Daboecia	1	Forsythia	6
Aronia	3	Cephalanthus	1	Dacrycarpus	1	Fortunearia	1
Arundo	1	Cephalotaxus	3	Dacrydium	4	Fothergilla	1
Asimina	1	Ceratonia	1	Dahlia	2	Franklinia	1
Asystasia	1	Ceratopetalum	2	Dais	1	Fraxinus	34
Athrosperma	1	Ceratostigma	3	Daphne	10	Freemontodendron	1
Athrotaxus	1	Cercidiphyllum	3	Daphniphyllum	2	Fuchsia	3
Aucuba	3	Cercis	9	Davidia	2	Furcraea	1
Azalea	3	Cestrum	2	Debregeasia	1		
Azara	6	Chaenomeles	7	Decaisnea	1	Gahnia	1
		Chamaecyparis	36	Decumaria	1	Garrya	3
Baeckia	1	Chamaerops	1	Dendropanax	1	Garuga	1
Banksia	4	Chimonanthus	11	Desfontainea	1	Gelsemium	1
Baptisia	1	Chionanthus	2	Deutzia	14	Geniostoma	1
Beilschmeidia	2	Chiranthodendron	1	Dichotomanthes	1	Genista	1
Berberidopsis	1	Choerospondias	1	Dichroa	1	Gevuina	1
Berberis	27	Choisya	2	Dicksonia	2	Ginkgo	1
Beschorneria	1	Chordospartium	1	Diervilla	2	Gleditsia	14
Betula	51	Chorisia	1	Diospyros	4	Glochidion	2
Bolusanthus	1	Chrysolepis	1	Dipelta	2	Glyptostrobus	1
Bomarea	2	Cinnamomum	1	Dipteronia	1	Goodia	1
Bowkeria	1	Cissus	2	Disanthus	1	Gordonia	3
Brachychiton	2	Cistus	11	Distylium	1	Grevillea	7
Brachyglottis	2	Citrus	3	Docynia	1	Greyia	1

Genus	No.	Genus	No.	Genus	No.	Genus	No.
Griselinia	2	Luehea	1	Phyllostachys	2	Sinojackia	2
Gymnocladus	2	Luma	1	Physocarpus	7	Skimmia	2
		Lycianthes	1	Phytolacca	1	Smilax	1
Hagenia	1	Lyonia	1	Picconia	1	Sophora	11
Halesia	3	Lyonothamnus	1	Picea	22	Sorbaria	5
Halleria	1			Picrasma	1	x Sorbomalus	1
Haloragis	1	Maackia	1	Pieris	3	x Sorbopyrus	1
Hamamelis	4	Macadamia	1	Pimelea	1	Sorbus	49
Hardenbergia	1	Machilus	2	Pinus	108	Sparmannia	1
Harpephyllum	1	Maclura	1	Piptanthus	1	Spartium	1
Hebe	5	Macropiper	1	Pistacia	2	Spiraea	21
Hedycarya	1	Maesa	2	Pithecellobium	1	Stachyurus	4
Hedysarum	1	Magnolia	90	Pittosporum	12	Staphylea	6
Helianthemum	1	Mahonia	7	Plagianthus	1	Stauntonia	1
Helichrysum	4	Mallotus	4	Planchonella	1	Stephanandra	2
Heptacodium	1	Malus	68	Platanus	9	Stewartia	6
Heteromeles	1	Mandevilla	1	Platycarya	1	Strelitzia	1
Heteromorpha	2	Manglietia	6	Platycladus	2	Strobilanthes	1
Heteropyxis	1	Maytenus	1	Plectranthus	4	Styrax	6
Heynia	1	Melaleuca	7	Podocarpus	14	Sutherlandia	1
Hibiscus	4	Melia	3	Podranea	1	Sycopsis	1
Hippophae	1	Melicope	1	Poliothyrsis	1	Symphoricarpos	2
Hoheria	3	Melicytus	2	Polygala	1	Symplocus	3
Holboellia	1	Meliosma	3	Polygonatum	1	Syncarpia	1
Hovenia	2	Merremia	1	Pomaderris	2	Syringa	15
Hydrangea	13	Metasequoia	1	Poncirus	1	Syzygium	2
Hymenanthera	8	Metrosideros	1	Populus	33		
Hymenosporum	1	Michelia	8	Potentilla	3	Tabebuia	1
Hypericum	8	Microbiota	1	Prostanthera	2	Taiwania	2
		Milletia	1	Protea	3	Talauma	1
Idesia	1	Mimulus	1	Prumnopitys	3	Tamarix	1
Ilex	40	Morus	4	Prunus	105	Taxodium	5
Illicium	5	Mrysine	1	Pseudocydonia	1	Taxus	9
Indigofera	7	Muehlenbeckia	3	Pseudolarix	1	Tecoma	1
Itea	5	Myoporum	1	Pseudopanax	4	Tecomanthe	1
Itoa	1	Myrica	3	Pseudotsuga	5	Telopea	1
		Myrsine	2	Ptelea	1	Ternstroemia	1
Jacaranda	1	Myrtus	4	Pterocarya	6	Tetracentron	1
Jasminium	7			Pterostyrax	3	Tetrastigma	1
Juglans	8	Nandina	2	Punica	5	Teucrium	3
Juniperus	41	Neillia	1	Puya	1	Thuja	10
		Neocinnamomum	1	Pyracantha	2	Thujopsis	3
Kadsura	1	Neolitsea	2	Pyrus	8	Tibouchina	1
Kalmia	1	Nepeta	1			Tilia	21
Kalopanax	1	Nerium	6	Quercus	123	Tipuana	1
Kennedia	2	Nestigis	2	Quillaja	1	Toona	1
Kerria	2	Nolina	1			Torreya	2
Keteleeria	3	Nothofagus	10	Rapanea	1	Toxicodendron	1
Kierengeshoma	1	Nuxia	1	Raphiolepis	2	Trachelospermum	2
Knightia	1	Nyssa	3	Rehderodendron	1	Trachycarpus	2
Kniphofia	3			Reinwardtia	1	Tripterygium	3
Koelreuteria	4	Ochna	1	Rhabdothamnus	1	Tristaniopsis	1
Kolkwitzia	1	Olea	2	Rhamnus	6	Trochodendron	1
Kunzea	1	Olearia	3	Rhaphithamnus	1	Tsuga	8
		Osmanthus	7	Rhodochiton	1	Turpina	1
Laburnocytisus	1	Osmaronia	1	Rhododendron	119	Tutcheria	1
Laburnum	4	Osteomeles	1	Rhodoleia	1		
Lagerstroemia	6	Osteospermum	1	Rhodotypos	1	Ugni	1
Lagunaria	1	Ostrya	3	Rhopalostylis	1	Ulmus	23
Lantana	1	Oxydendrum	1	Rhus	9	Umbellularia	1
Lardizabala	1	Ozothamnus	1	Ribes	6		
Larix	6			Ricinus	1	Vaccinium	1
Laurus	4	Pachysandra	1	Robinia	3	Vallea	1
Lavandula	6	Paeonia	5	Romneya	1	Viburnum	29
Lavatera	1	Paliurus	1	Rosa	11	Vitex	2
Ledum	1	Pandorea	1	Rosmarinus	5	Vitis	6
Leonotis	1	Parakmeria	1	Rubus	4		
Leptospermum	5	Parapiptadenia	1	Ruscus	1	Weigela	7
Lespedeza	3	Parasyringa	1	Russelia	1	Weinmannia	2
Leucadendron	2	Paratrophis	1			Widdringtonia	3
Leucaena	1	Parrotia	1	Salix	18	Wisteria	9
Leucothoe	2	Parrotiopsis	1	Salvia	11		
Libocedrus	2	X Parrotisyopsis	1	Sambucus	2	Xanthoceras	1
Ligularia	1	Parthenocissus	2	Sapindus	1		
Ligustrum	8	Paulownia	7	Sapium	2	Yucca	1
Lindera	6	Pennantia	1	Sarcococca	4		
Lippia	1	Perovskia	1	Sassafras	2	Zanthoxylum	8
Liquidambar	6	Persea	1	Saxegothaea	1	Zauschneria	1
Liriodendron	4	Petteria	1	Schima	4	Zelkova	5
Lithocarpus	6	Phaedranthus	1	Schinus	2		
Lithospermum	1	Phellodendron	5	Schisandra	3		
Litsea	2	Philadelphus	16	Schizophragma	3		
Livistona	1	Phlomis	2	Sciadopitys	1		
Lomatia	5	Phoebe	1	Senecio	2		
Lonicera	19	Phoenix	1	Sequoia	5		
Lophomyrtus	2	Phormium	5	Sequoiadendron	1		
Loropetalum	1	Photinia	12	Sesbania	1		
Luculia	1	Phyllocladus	3	Sibiraea	1		

Total of 570 genera

11 Formation of the Trust

There is a fascinating link in the Eastwoodhill story between a country hotel in Cornwall in the year 1930 and a campaign waged by Bill Williams, in the late 1960s and 70s, to establish a charitable trust in order to preserve the Eastwoodhill arboretum. It is a link which involves a young Englishman on his honeymoon and a chance meeting with Lady Thistleton-Dyer.

The honeymooner was Bill Way, then aged 34. He and his wife Aileen were placed at the same table as Lady Thistleton-Dyer in the dining-room. The conversation during their first dinner together turned to plants and gardens – not because Bill Way had any particular interest in botany at the time, but because Lady Thistleton-Dyer was both the daughter and the widow of former directors of the Royal Botanic Gardens, Kew. Her father was Sir Joseph Dalton Hooker, the director from 1865 to 1885; her late husband, Sir William Turner Thistleton-Dyer, had been her father's deputy for 10 years and served as director himself from 1885 to 1905. The book Royal Botanic Gardens Kew: Gardens For Science and Pleasure credits Sir William with having introduced "with some reservations" the first women gardeners at Kew in 1896. To ensure that their presence did not distract their male colleagues they had to wear a knickerbocker suit, thick woollen stockings and "a most unbecoming peaked cap".

More than 60 years after that meeting in Cornwall, at the age of 97, Bill Way could still recall how absorbed he became with Lady Thistleton-Dyer's conversations about the role of Kew in the development of the British Empire through the propagation of rubber and tea plants and through quinine (from the bark of a species of Cinchona) as treatment for malaria. That her stories had an impact on his mind was evident in 1970 during his public relations campaign of support for an Eastwoodhill trust. He wrote: "It is interesting to note how many gardens which today are of world renown have attained their prominence by performing a definite function. The outstanding example of this is, of course, Kew which, at the time when the British Empire was at its peak, was able to assist in the economic development of various territories by obtaining material to start industries such as tea and rubber." Using Kew as an example, he was arguing that Eastwoodhill would have a "definite function" in the future, preserving the quality of the environment, as a source of shrubs and trees to suit different climes.

In fact, Eastwoodhill had already made an impact in New Zealand. Several *Malus* species identified there had been used as rootstock enabling apple crops to resist the effects of fire blight.

Bill Way's global perspective stemmed from 30 years of

Bill Way (above) helped H.B. Williams in his efforts to secure Eastwoodhill's future. (Photo Dorothy Way)

employment with the international merchant firm of Jardine Matheson and Co. Ltd. As a young man he worked in the import department in Hong Kong and then transferred to Tsingtao on the China mainland where he learned to speak Mandarin. Based in Shanghai in 1941, he was interned by the Japanese. He busied himself teaching some of the American and British children in the camp – and taught them so well that their examinations, channelled through the Swiss Consulate, were recognised after the war in both Britain and the United States.

Bill settled in New Zealand, at Christchurch, where his wife died in 1950. In 1955 he married Dorothy Adams and moved to Gisborne, working in the insurance field until he retired in 1960. He celebrated his retirement by taking Dorothy on a trip to Britain – and a visit to Kew. Retirement for Bill meant throwing himself into various community causes. He became director of the Gisborne Museum and, in an honorary capacity, worked with the Historic Places Trust in its fight to preserve Captain James Cook's first landing place in New Zealand, on the Gisborne foreshore. He was honorary treasurer of the Farm Forestry Association's Gisborne branch from 1964 to 1967.

It was inevitable that Bill Way would become involved with Eastwoodhill. With his fervour, his concern for trees and plants and his business background, he was the right man in the right place at a time when Eastwoodhill was relatively unknown in New Zealand except among people with a specific interest in dendrology. There was little chance that charitable trust status as a means of funding the arboretum could be achieved unless the Government and its advisers recognised its true value. In that respect, Bill's passion for letter-writing, almost rivalling that of Douglas Cook, became an important weapon. He did not simply "write" letters; he drafted them, carefully weighing every phrase and sentence. Bill Williams welcomed his involvement – "He was a real help and was vitally interested."

The importance of Bob Berry's catalogue became clear: without it botanists within government departments could have no way of assessing the importance of the arboretum, not only nationally but internationally. Bill Way and Bob Berry worked in tandem, spreading the gospel.

Bob was secretary of the Farm Forestry Association's local branch and Jim Holdsworth, stud breeder and farm forester, was its chairman, when concern for Eastwoodhill surfaced at the association's New Zealand conference in Gisborne at the end of February, 1969. A group of the Gisborne members had decided to do without the paid clerical assistance usual when hosting a national conference and carried out the duties themselves. They raised $300 for the arboretum.

Bill Way joined the Royal New Zealand Institute of Horticulture specifically, it seems, so that he could attend with Bob Berry the annual conference in Napier in February, 1970 and push the case for an Eastwoodhill trust. The pair must have been persuasive: on a show of hands numerous delegates indicated that they would be prepared to follow the lead of one delegate who had pledged personally $50 a year to a fund for Eastwoodhill. Bill Way followed a similar tactic by arranging to become, along with Bob Berry, an affiliate member of the New Zealand Institute of Foresters. As soon as he had gained the new status, he journeyed to Rotorua for talks with senior staff of the Forest Research Institute.

Meanwhile Bill Williams, frustrated in his efforts to make the farm at Eastwoodhill sufficiently profitable to support the arboretum, had indicated to the Royal New Zealand Institute of Horticulture that his mother, Mrs H. B. Williams snr, was prepared to buy Eastwoodhill and gift it to the institute together with a $10,000 endowment. Its members were sympathetic. However, the stark reality was that the institute did not have the resources to maintain the arboretum. Thoughts of government assistance were dashed by the Minister of Lands, the Hon. Duncan McIntyre, who stated that neither the Forest Service nor the Department of Lands and Survey could take over the property, nor contribute anything other than the advice of its officers.

It became clear to Bill Williams and his legal advisers that the solution would lie in "the establishment of Eastwoodhill as a permanent institution with its own governing body and ability to receive financial donations which would be exempted from death duties if given in the lifetime of the donors or at their death". Bill Way was given authority in 1970 to assemble information to form part of a case to the Government. Authority for such an institution would require an Act of Parliament through a Private Member's Bill.

This was no quick solution. Further years of letter-writing and discussion were ahead. Bill Williams' lawyer, following the retirement of Colonel Gambrill, was Tom Thorp (later Justice Thorp). He wrote in 1971 to Phil Jew, supervisor of Parks and Reserves for the Auckland Regional Authority, asking if he would be prepared to undertake an investigation of Eastwoodhill to establish its botanic value. Phil accepted: he would conduct the study during annual leave from the ARA, with no charge for his time. Phil began his task with a visit to Gisborne, spending the weekend of August 14-15 as a guest of Bill and Dorothy Way. His report was delivered to Tom Thorp at the end of April. He noted, in his introduction, that some of the people who had known Eastwoodhill in former years, and with whom he had held discussions, had reservations as to whether an arboretum was within the resources of a provincial centre.

"While I can well understand their reaction," he observed, "I am encouraged to take a more positive attitude by the ability of smaller communities to achieve goals which would be virtually impossible in larger centres." He considered that "the development of a public arboretum at Eastwoodhill has to be considered justified. "

There was a sign of hope for Bill Williams and his team when the Auckland district council of the RNZIH reported that one of their group who had visited the arboretum had written to the Minister of the Environment, the Hon. Joe Walding, suggesting that the arboretum be preserved by including it in the national parks system. The Minister's reply indicated that the Government was sympathetic and might give some assistance "when the time comes". National Park status was not what the group sought, but the positive reaction of the Minister was cause for optimism.

Breakthrough came in mid-1974 when a draft for a Private Member's Bill, prepared by Tom Thorp, was approved both by the government departments involved and by the Minister for the Environment. The bill, dated June 13, 1975, was presented by the Member of Parliament for Gisborne, Trevor Davey, and was passed.

The Act provided for a board of five members – a chairman appointed by the Governor-General and one member to be appointed by each of the following: the Cook County Council, Gisborne City Council, the Poverty Bay Horticultural Society and the Gisborne branch of the New Zealand Farm Forestry Association. The Act also established an advisory committee of three members to be appointed by the Director-General of the Department of Scientific and Industrial Research, the Council of Massey University and the New Zealand Forest Research Institute.

Of crucial importance was the provision, occupying only two lines of the Act as "clause 16", stating: "It is hereby declared that the purposes of the Board are charitable purposes." Bill Williams vested the Eastwoodhill property, buildings and stock in the trust, and the first gift to the board's endowment fund, of $50,000, was from the M.A. Williams Charitable Trust established by his mother.

Bill was appointed chairman and the first members of the

trust board were Terence Williams, chairman of the Cook County Council; Brian Davis, the City Council's superintendent of Reserves; Bob Berry, appointed by the Horticultural Society; and Fred Faulkner of the Farm Forestry Association. Together they represented a wealth of horticultural and administrative experience.

Terence was a distant relative of Bill through the two missionary brothers who arrived in New Zealand in the 19th century. Terence was descended from the elder brother, Archdeacon Henry Williams; Bill was the great-grandson of William. Terence's father, Henry Carleton Williams, settled in the Gisborne district in 1905 and developed, with his brother Claud, a farm at Muriwai, south of Gisborne, called Sherwood which originally was part of Maraetaha Station. Brought up on the property, Terence managed it on behalf of his father and uncle after World War 2. He inherited and developed not only farming expertise but also a respect for trees and shrubs with which the property was well-endowed.

Fred Faulkner was a neighbour of Terence at Muriwai. His father, Albert Faulkner, built his homestead there on a hillsite overlooking Poverty Bay. As Wairakaia, it was to become a garden haven known throughout the region and beyond. Its garden, which was developed in stages, was almost 90 years in the making. Fred, and his wife Mary, concentrated on a formal garden around

On the occasion of the unveiling of the IDS plaque in 1978, trust board members gather under the lagerstroemias in the Homestead Garden.
From left are Bob Berry, John Overbye (later the Farm Forestry representative), H.B. Williams, Terence Williams, Bob Bell (MP for Gisborne), Dinah McIntyre and the Hon. Duncan McIntyre (Minister of Lands), Dan Weatherall (Curator), J.P. Salinger (Eastwoodhill Advisory Committee representing Horticulture Department of Massey University), Fred Faulkner, Brian Davis, Trevor Cook (Trust Board secretary), R.M. Greenwood (Eastwoodhill Advisory Committee representing the Department of Scientific and Industrial Research). (Photo Dunstan and Kinge)

Joe Oates has acted as treasurer for the Eastwoodhill Trust Board since its inception. (Photo McCullochs)

Bill Way (right) lent his support to Eastwoodhill right to the end of his days and faithfully attended all the Friends open days. He is pictured here with H.B. Williams on the occasion of the visit of David Bellamy in February 1989. (Photo Dorothy Way)

the house and built the first of the stone walls which became a feature of the tennis court gardens. In 1967 they retired and their son Rodney (later to become a trust board member) and his wife Sarah continued the garden's development.

Brian Davis, who was raised in Dannevirke, brought the skills of a nurseryman to the board table. Before joining the Gisborne City Council staff as assistant reserves superintendent in 1966 he had completed his horticultural apprenticeship at Wilson's Nursery in Hastings and had worked at Langbecker's Nursery in Bundaberg, Queensland.

Bob Berry, of course, was already deeply involved with the arboretum.

The first meeting of the board on August 13, 1975 was attended by Cliff Costello, commissioner of Crown Lands, who said that his organisation would be happy to provide secretarial services through Trevor Cook of the Lands and Survey Department. Joe Oates, of the accountancy firm McCulloch, Butler and Spence, was appointed treasurer. Joe had been involved with the Williams family businesses for many years and had been a behind-the-scenes player throughout the quest for the formation of the trust. It was a relatively short meeting, lasting an hour and 20 minutes. But it was an historic one.

Bill Way, the trusty "unofficial spokesman", never became a board member, having passed the age limit of 70 laid down in the Act, but he attended numerous meetings over the years by invitation. Just as Douglas Cook had dreaded the effects of a nuclear war, Bill became increasingly concerned about the effects of acid rain on plant life in the Northern Hemisphere. He studied countless articles on the subject from throughout the world. In a letter to Graham Weston in 1988, he quoted a report in the British journal New Scientist outlining the damage to trees on the mountains and in the lake districts of Scandinavia – "the loss of trees had caused massive soil erosion and frequently the death of villagers". When he learned that Graham was planning a visit to Britain he suggested to the trust board that his friend could discuss with European botanical institutions the board's willingness to act as a gene bank for seed. The board took his advice and wrote to Graham on the subject. In the event, Graham had to postpone the visit that year, but Bill Williams followed up the idea during a trip to Britain with a meeting at the Royal Botanic Gardens, Kew.

12 The Clean-up Years

Dan and Molly Weatherall beside the Sundial Circle. Dan's almost-silent little Honda 90 farm motorbike surprised many a visitor walking in the arboretum. Molly was a generous hostess in and about the garden and gave a personal welcome to most of the visitors in those days. (Photo Dennis Weatherall)

There is a tree beside Gisborne's Gladstone Road, at the corner of Herbert Road – a dwarf weeping peach which, in the springtime, has caused many a pedestrian to stop and stare. It is among many plants grown in the city from Eastwoodhill seed or cuttings, as Douglas Cook had been generous over the years with plant material. What makes the peach unusual is that it flowers in two colours, pink and white. Some branches produce blossoms of both colours; other branches are all-white or all-pink. This characteristic "comes true" from seed.

The tree was nurtured in a pot at the Eastwoodhill house by Molly Weatherall who, with her husband Dan, cared for the farm, the house and – as well as they could – the arboretum from 1974, a year before the trust board was established, until 1982. Molly was so enchanted by the peach that she made inquiries about its origin. She learned that the original seedling had been included in a consignment of plant material imported by Douglas Cook. He had inquired too about the peach seedling because the consignment list made no mention of it. But the nursery concerned

Dan Weatherall (right) was still at Eastwoodhill as preparations were made for the building of the pavilion. Here he and Terence Williams attempt something with the 'old Fergie' though obviously not successfully as the stump is still there. (Photo Terence Williams)

A Friends of Eastwoodhill membership card issued during Molly Weatherall's time as secretary. The blacking out covers the year 1978 and the old phone number 599.

replied that it had no record of the seedling. At the publication date of this book the tree remains unnamed, its origin unknown.

Perhaps the peach could be regarded as a living memorial to the Weatheralls' eight years of toil, against difficult odds, at Eastwoodhill. When Dan was appointed to the position of farm manager, Bill and Jo Crooks still occupied the house. Bill was now growing weary after the years of unrelenting service and, while tending the sheep and cattle, could not cope with the demands of the arboretum. Before Douglas Cook's death, continued planting had taken precedence over maintenance: weeds, creepers and self-sown seedlings had begun to take hold. The Weatheralls, however, were hardy, resilient folk from deep in the South Island. They were used to challenges.

Dan, who was born in Central Otago, had worked on the Bluff oyster boats during the season and then, like most of the fishermen, went shearing. From the time he and Molly were married they were always close to the land. After growing potatoes, then raspberries, they had a sheep and cattle farm. In the 1960s they bought a farm near Te Puia on the East Coast, eventually moving to Gisborne – perhaps to retire. But Dan was becoming bored when Bill Williams placed a newspaper advertisement seeking a farm manager for Eastwoodhill. Dan applied and got the job. At the time the couple had planned a trip to Britain and the Continent. Bill Williams allowed them the three months to take their trip. Molly said later: "We made a point during our travels in Britain and on the Continent to visit some arboreta. We were amazed at what was involved in caring for such a place. I thought to myself, what have we got ourselves into?"

She may well have wondered. It rained at least part of each day for 27 days from the date they moved in. The paths were overgrown. They had no map or plan of the arboretum's layout and twice, as they attempted to explore, they became lost trying to find their way back to the house. Molly thought the house itself "rather weird" with its passage of two different widths. They had bought a table from Bill Crooks. Because of that passage Dan had to remove a window frame and manhandle the table out of one window and in another in order to position it in their dining room.

Despite the state of the arboretum, the priority for Dan had to be the farm. Bill Williams, who was trying to make the property an economic unit, needed to maintain and hopefully increase income from the sheep. Any serious onslaught on the weeds, undergrowth and general rubbish in the arboretum had to wait. Molly, however, looked around the garden and wondered where to start. ("I decided to begin at the veranda and work my way outwards"). Camellias were growing out of control and shutting out the light from plants below. She had to enlist Dan's help to top many of the camellias in the homestead garden by as much as 20ft. *Ailanthus altissima* had gone wild, springing up like weeds and forming virtually a solid mass "you could have walked on". Molly must have smiled ruefully about the plant's popular name of Tree of Heaven. She regarded it as being "a sod to seed". After Dan cut down scores of them they resorted to ring-barking and poison for control.

In tending the house garden, Molly made her own selections of what to plant but never pulled out any existing material. ("We never moved without consulting Bob Berry. He was extremely helpful and very knowledgeable. We would have been sunk without him.") Because visitors continued to arrive to see the arboretum it was essential to clear the pathways. At least the sheep made a contribution to the appearance of the property by keeping the grass down. To visitors, Dan would refer to them jokingly as his "mowers".

After the Eastwoodhill Trust Board was established on August 13, 1975 its members saw an opportunity to formalise a sort of club of enthusiasts as a means of sustaining visitors' interest, of building support, gaining donations, and of providing some guidelines for visits which had been on a rather random, uncontrolled basis. The board appointed Molly honorary secretary and had membership cards printed. The organisation was called Friends of Eastwoodhill. Molly would record the names of visitors to the house, noting any donations beside their names. The money was passed on to the board's accountants. She not only took great pains to make visitors welcome, but impressed upon them, over chat and a cup of tea, what Eastwoodhill meant as a horticultural treasure.

There was growing concern about the backlog of clearing work. Brian Davis, then parks superintendent of the Gisborne City Council and a board member, recalled: "It could have taken six men working fulltime for a couple of years to clear it." Another board member, Fred Faulkner, paid numerous visits to Eastwoodhill and, with a spray-pack on his back and a spade at his side, would attack the weeds.

Help came in the form of the PEP scheme (short for Project Employment Programme), which was administered by the Department of Labour in association with local authorities, to provide useful work for the unemployed. Through Brian Davis, the board was able to engage three men. At that time Christopher Williams, a son of Terence, was working in the council parks

Chris Williams was commissioned by the trust board to produce its first management plan. (Photo Gisborne Herald)

department as summer vacation employment while he was halfway through a two-year Diploma in Horticulture course at Lincoln College. Brian arranged for him to drive the PEP men to and from Eastwoodhill each day, and Chris worked with them.

He became deeply interested in the arboretum. As part of his diploma he wrote a 40-page dissertation on Eastwoodhill which was printed in 1980. Because of that, when the board decided to have a management plan prepared, they commissioned Christopher to do it. He produced a draft plan in October, 1984, for public discussion and input and it was adopted, three years later, as the management plan. A few years later Chris extended his qualifications by gaining a Bachelor of Landscape Architecture degree at the Royal Melbourne Institute of Technology. By presenting the Eastwoodhill Management Plan he was granted a pass, without further study, in the subject of management planning.

Molly Weatherall estimated that over the eight years of their tenure, Dan had removed almost 500 unwanted saplings "twice as tall as himself". She described the period as "a learning experience" but added: "We had had enough of battling with Eastwoodhill. By 1982 it was time to retire and to pursue hobby interests. They vacated the house and retired to Herbert Road where Molly planted that weeping peach in the front lawn.

Two important decisions were made by the trust board on the retirement of Dan Weatherall. One was to appoint Warwick Spence to manage the farm. The other was to give the board more direct control over activities on the property by appointing one of its members as supervisor of the arboretum. That first supervisor was Terence Williams who had taken a close interest in the park from the day the board was established.

Terence brought to the position a reputation throughout the district for tact and diplomacy. As a long-serving councillor then chairman of Cook County Council he had frequently defused tensions between members representing hill-country farmers and those on the Flats. When a dispute arose between the County and the City Council over land for a rubbish tip, he dampened down newspaper controversy about it by ringing the Mayor, Harry (later Sir Harry) Barker, suggesting that they should settle their differences over the telephone, or in private discussion, rather than in the pages of the Gisborne Herald.

When a new curator, Kevin Boyce, was appointed, Terence was determined to push ahead with clearance in the arboretum as expeditiously as possible, knowing that new plantings would have to wait until the huge task was complete. There was much to be done. Despite the best efforts of the Weatheralls, some areas were totally overgrown, smothering important species of plants; self-sown weeds abounded; areas were shaded almost to the point of darkness. Kevin recalled: "To start with, we had to tackle the general impenetrability of the place."

In 1985 several large trees were removed from Three Bridges area giving light to rhododendrons and deciduous shrubs. From left, arboretum curator Kevin Boyce, garden group members Dawn Jefferd, Mary Bush and trust board supervisor Terence Williams. (Photo Gisborne Herald)

Unlike Dan Weatherall, Kevin Boyce had no duties related to the farm. He now occupied the house with his wife Anne and their children and brought to his task a background of five years as groundsman at Cook Hospital which was then Gisborne's public hospital, set on the top of a hill. The property incorporated a large number of trees. He had acquired experience in working alongside PEP workers, many of them Maori, and he understood their cultural attitudes. Because he had a good head for heights, he was able to establish from the outset that if any tree had to be climbed, he would do it. It was a form of platform pruning with a jacksaw. He first climbed a 24ft extension ladder before negotiating the first branch, worked his way up, then pruned from the top branches down. His prowess won him mana (prestige) in the eyes of the men.

He began with a five-man gang and a supervisor. Skill was needed in felling a tree, guiding it precisely in the direction of its fall so that it would not damage valuable trees around it. Bob Berry was a frequent visitor, advising on the trees to fell and those to preserve, and indicating the location of half hidden, valuable plants. Terence Williams called on a regular basis to discuss progress.

The carpentry class of the Tairawhiti Community College began construction of the visitor's pavilion in the garden. The shed adjacent was being used by curator Dan Weatherall as his wood shed and garage and to tie up his farm dogs. This same building was where the Cooks lived after the 1931 fire (see page 26) and had previously been located where the Douglas Cook Centre now stands. (Photos Terence Williams)

In the early days of the project Kevin noticed that some of the men would bring a packed lunch, others would not. Such a situation under Maori protocol demanded that those with food share it with the others. In the physically tough work conditions, he was concerned about the effect on the men, as the afternoon drew on, of inadequate lunchtime nourishment. It could even lead to lapses in safety standards as well as efficiency. With his encouragement the men set up a tripod cooking arrangement and two camp ovens on the flat area of land that later was to become the site of the Douglas Cook Centre.

A severe drought throughout the region in the summer of 1982-83 created problems. Douglas Cook's doubts about the suitability of the property for rhododendrons were borne out. The men toiled to keep the water pumps going but the rhododendrons began to wither. Twenty-six species were lost.

With visitor numbers increasing, the board decided to provide toilet facilities and shelter for people to have tea and relax. A site for a pavilion was cleared on a mound off the west end of the top lawn and some Lawson cypress were removed. The simple but effective design was provided in part by an Auckland architect and modified by Derek Phillips of Gisborne. He provided working drawings and assisted greatly at no charge to the board. All the *Cupressus macrocarpa* and *Cupressus torulosa* timber for the building was given by several farmers; generous discounts were given by the sawmiller and the suppliers of the roofing iron. Gifts of windows, doors and interior fittings further reduced the cost.

An arrangement with the Tairawhiti Community College carpentry school provided a young and eager workforce. The class of 10 apprentices, under the supervision of their tutor Philip Gaukrodger, cut and assembled the framing at the school. Then they boxed and laid the concrete floor at Eastwoodhill, and erected the building. Because the students had only a limited timespan to devote to the task, a professional builder attended to the finishing work. Win Ellis, an electrical contractor, attended to the electrical installation without charge. A working bee, spreading their efforts over several days, laid the brick surrounds and the approach path. They were helped by a number of retired men under the voluntary guidance of a professional block-and-brick man, David Aitken.

The pavilion had been the first material development at Eastwoodhill since the declining years of Douglas Cook, and the board celebrated with an official opening. All the donors of money and material, the carpentry team and the PEP workers were among the guests for the grand occasion on a sunny Saturday, April 30, 1983. The offical opening was performed by Bryan W. Robinson who had flown from Auckland with his wife, with Dennis Yates and Mrs Yates, representing the Selwyn Robinson Trust of Auckland – one of the principal donors of special funds.

Besides the PEP workers, there were other helpers at the arboretum. People came to know them as "the gardening girls". Every Thursday morning at 8.30 they would arrive at Eastwoodhill with their forks, trowels and spades, and a cut lunch, to attack the weeds and tend the flower beds around the house. Weary, a little grubby and sometimes soaked from rain, they would leave for home at 4.30 and sink gratefully into a hot bath before cooking the evening meal for their families. It was a self-imposed labour of love.

Dawn Jefferd, Mary Bush, Robin McIldowie and Bev Bridge were farmers' wives who shared not only an enthusiasm for gardening but also concern for the health of Eastwoodhill. Their efforts inspired others, including their husbands. The idea of a community effort to assist the arboretum snowballed and they were to become a moving force in the revitalisation of The Friends. Bev travelled to Eastwoodhill from Manutuke, Dawn and Robin lived on the Totangi Road in Ngatapa and Mary was practically a neighbour of the arboretum, within strolling distance along Wharekopae Road.

Mary and her husband Spencer had known Douglas Cook from 1960 when they settled on their farm. Spencer's father Thomas Bush, like Douglas, had been a Gallipoli man. Douglas invited them to visit for afternoon tea on a Sunday. At his suggestion, Mary baked a sultana cake for the occasion. Douglas escorted them around the arboretum and pointed out different species of trees. Before they left, "Mr Cook" gave them seeds of a trident maple which Spencer planted. And he became a "tree man" too, later serving on the Friends committee and as a supervisor of Eastwoodhill. Many a wooded property in Ngatapa bears testimony to the fact that Douglas Cook never missed an opportunity to spread "the gospel of the trees".

The idea of the "gardening girls" group began to sprout when Dawn was chatting to Terence Williams about the state of the Eastwoodhill house gardens. Help was certainly needed, but the board had to be wary, bearing in mind the valuable species that Douglas Cook had planted. Many important plants were now totally obscured by weeds. Haphazard, over-enthusiastic weeding could have been disastrous.

Dawn discussed the problem with Mary and they developed the idea of forming a garden group with Bev and Robin. On the advice of Terence, Dawn wrote a formal letter to the board and

Mr H.B. Williams and his wife Elizabeth about to enter the pavilion on the occasion of its official opening on April 30, 1983.

Arboretum curator Kevin Boyce and the PEP workers lifted sandstone rocks from the overgrown south-facing rockery that Douglas Cook had established in the late-1920s to build rock walls and establish gardens around the pavilion.

gained approval for their project, provided it was confined to the garden and did not encroach into the arboretum. In June, 1984, they began in the area of the newly-erected pavilion. At the start, the gardens were a wilderness: in Dawn's words, "everything was growing everywhere".

Terence Williams, a trust board member, arboretum supervisor and strong supporter, was more commonly seen in working garb at Eastwoodhill. Here he stands beside the sundial which he had fixed to the plinth for the first time since 1926. (Photo Garry Clapperton)

It was careful, painstaking work in order to avoid damage to valuable plants. But there were delightful surprises too – "We were never sure what was going to pop up, there were so many bulbs." They consulted gardening books to identify some of the plants. A friend even wrote to South Africa in an effort to identify one flower, but without success. When there was heavy work, beyond the women's muscle-power, they enlisted their husbands – Spencer Bush, Tiny Jefferd, Jim McIldowie and Hugh Bridge. The men never hesitated. Terence, as supervisor, was always involved. For some major tasks, like the re-forming of Bog Garden, for carting in new soil and for shifting fence lines, Robin and Jim called in some of their men and a truck from their Totangi Station.

As the work progressed, the group became short of plants. Dawn wrote to the Pukeiti Rhododendron Trust in Taranaki for help and received a positive response from the curator, Graham Smith, with an offer of rhododendrons and azaleas, provided that someone came to get them. Robin and Dawn set off in a small car and returned with the back seat crammed with hundreds of dollars worth of material. Later the group obtained plants from the Auckland Regional Botanic Gardens in Manurewa.

While the gardening continued, another topic cropped up in discussion with Terence. There was a clear need for wide-based

Stephen Jones

The Douglas Cook Centre.

Stephen Jones

The Garden pavilion

Garry Clapperton

Stephen Jones

The ixia bed at the end of Palm Terrace.

Garry Clapperton

Herbaceous border on the Sundial Drive.

The Homestead Garden in 1996

Hostas in the ponderosa lemon bed

Stephen Jones

Mollis azaleas in the Sundial Circle.

Garry Clapperton

Cannas and herbaceous beds (above) were a feature of this part of the Homestead Garden in 1961. A weeping beech Fagus sylvatica 'Pendula' *shows at upper left. But today…*

…the weeping beech is the main feature of the Lower Lawn, a simpler treatment of the area.

Garry Clapperton

Stake and tie for a long summer display of delphiniums.

Garry Clapperton

Spraxias along the Palm Terrace.

Garry Clapperton

xxviii

community support for the arboretum. Terence regretted that the original Friends of Eastwoodhill had flagged and was now little more than a means of recording donations by some generous visitors. Terence and his wife Joyce invited the "girls" and their husbands to a meeting at their home in Wainui. They agreed that they should visit Pukeiti to study the way it operated there. All 10 of them travelled, more or less in convoy, to Taranaki. On their return, Spencer wrote to the board setting out suggestions for a formally constituted association. That was to open another new era in the Eastwoodhill story . . .

The Homestead Garden was to benefit through the next decade, and beyond, by the involvement of a wider group of gardening ladies. Nine garden areas were allocated to individual ladies who either continued to look after their area alone, or else brought in extra volunteers to assist them. The numbers grew to a total of 20 volunteers working on a regular basis with a further seven attending three to four times a year on an assigned "major project day". The amount of time each lady could give varied according to family commitments. No pressure was exerted on them. The emphasis from within the group was to make their time spent at Eastwoodhill an enjoyable experience, not an onerous one. The overall result has been a consistent advance in the level of presentation of the Homestead Garden.

As well as maintaining their own areas in the garden, the volunteer garden ladies frequently call for extra help on special working bees to tackle other areas in the Homestead Garden.

Another major advance in the garden story came in October, 1992 when the trust board appointed Gordon Collier, of Titoki Point garden near Taihape, to be responsible for overall design of the

Homestead Garden area. He immediately made some major changes in the layout, including removal of several paths in the confused network that Douglas Cook had laid out. Less choice of paths made it more straightforward for visitors. Gordon Collier also gave overall guidance as to the type of material to be used in different areas, and to the choice of permanent structural and background material. But he left it to the ladies to make their own choices as to flowers and colours.

From within the garden group came a dedicated commitment to the production of plant material, mainly with the assistance of the nursery at the Rural Studies Unit of Tairawhiti Polytechnic. Local and overseas catalogues were scoured as inspiration for new annuals and perennials that would fit the parameters set for each area. Over the years some limited expansion of the garden area has been made, but overall only part of the Homestead Garden is to be worked intensively; other areas will be less intensive and planted in ground covers and shrubs.

The initial thought of keeping the garden in "the style of Douglas Cook" was set aside. On reflection, Douglas had no style. When a long-time friend of his was asked if the garden was ever tidy she replied with a smile: "Yes, the bit that he was working on!" The garden has continued to evolve and is becoming a major drawcard in the attractions of Eastwoodhill. It is the only major "public" garden in Gisborne.

In the lower garden, a superbly-finished concrete wall built by Douglas Cook stood for long years with no soil or planting behind it. In 1985, the garden group began the long overdue finishing touches. Here Hugh Bridge shifts camellias from another part of the garden and adds soil. (Photo Friends of Eastwoodhill)

13 A Young Botanist

J ust as Douglas Cook had a dream of trees and the future, so did a Havelock North lad called Garry Clapperton, later to become curator of Eastwoodhill. He was a seventh former at Karamu High School in Hastings when he read A Garden Century, a history of Christchurch Botanic Gardens. He was also reading in class at the time of European forests, of hardwoods, oaks and birches. Years later he recalled, "I had a dream of being involved in a forest made up of exotic species." That vision was to influence his career path throughout his youthful years until he found fulfilment among the tall trees at Ngatapa in August, 1985.

He was born in Hastings in 1953, the son of a carpenter, Don Clapperton, who was to become a frequent visitor to Eastwoodhill over the years, volunteering his carpentry skills to many projects. Garry seems to have inherited plantsmanship in his genes: there were professional gardeners back through generations on the paternal side of the family. As a 12-year-old, he began to look after a half-acre property, planting, pruning and weeding for Lynne Warren. He was also a regular visitor to his "green-fingered" grandmother to admire her variety of plants.

At the age of 14, Garry became a member of the Hastings/ Havelock North branch of the Royal Forest and Bird Protection Society. More remarkably, considering his age, he was seconded within weeks to its committee. During New Zealand's first national Conservation Week in 1969 the teenager initiated a roadside flax planting project at a memorial between Napier and Clive. The area was marked as the site of a mission station founded by the pioneer missionary William Colenso. With the aid of a truck supplied by the Ministry of Works and a party of Boy Scouts he ventured into the Kaweka Forest to collect native mountain flax, *Phormium colensoi* (later renamed *Phormium cookianum*). It was believed to have been the first roadside planting of native flax in New Zealand. In his Forest and Bird Society role, Garry was in the company of a group of amateur lady botanists on trips, and from them had his introduction to native plants and the use of botanical names – "It was a very positive influence," he recalls.

He had a well-rounded boyhood: a talented sprinter consistently winning championships and setting records; he also learned to play the piano. But botany was his greatest joy.

When he completed his seventh form year it seemed a logical step to join the New Zealand Forest Service – a chance to pursue a career among the trees. During holidays he had worked in forests and had been involved in a beech forest survey. Employment with the service provided him with the opportunity to study at Canterbury University for a Bachelor of Forestry Science

Curator Garry Clapperton (left) and David Bellamy head towards the Daffodil Patch on the occasion of Professor Bellamy's visit in 1989. (Photo Gisborne Herald)

Eastwoodhill has no permanent natural water . To enlarge on the storage system built with a Stanley Smith Trust grant, the trust board built a one million gallon capacity dam on Big Hill in 1992. It is near the arboretum's highest point of 297 metres above sea level and provides a gravity-fed water supply. The photograph shows it in 1993 as it was filling. The farmed area beyond the dam is part of the original Eastwoodhill. (Photo Terence Williams)

degree, a course which he later dropped in favour of a B.Sc. in Botany. At his introductory course at the service's training centre at Whakarewarewa, Rotorua he was one of 80 trainees, 20 of them forester trainees and the others ranger trainees. Garry achieved top marks for the intake.

His boyhood studies and experience with plants were of inestimable value both with the Forest Service training and at university. The outstanding example of that concerned the study of plant hormones. While still at school he had become fascinated by the subject and wrote to the ICI chemical company to obtain tablets of giberallic acid (plant hormone) in order to conduct his own experiments. He knew from his reading that dwarf peas and beans could be "turned back" into runners and that hormones could influence fruit, such as increasing the size of grapes on the vine. He must have smiled to himself when, at university, the subject of a biology course was plant structure, with emphasis on hormonal influences. Not surprisingly, he topped the course out of 250 students.

During those university years he established a friendship through gardening and plants with the matron of College House, Meriel Farnsworth, who owned a property with cottage gardens at Waddington in Central Canterbury. They talked for hours about plants and Meriel introduced Garry to a number of gardening books. On weekends he would stay on the property, tending the garden by day. He increased his knowledge of some of the more intricate gardening techniques.

The Homestead Garden had many uncompleted corners. A rockery planting was established in 1985 where, for many years, a concrete water trough had sat on the surface surrounded by ivy. The weeping peach is a seedling of the original that came unnamed from Japan in 1920. (Photo Friends of Eastwoodhill)

Disillusionment arose in his association with the Forest Service through its emphasis on radiata pine – an exotic tree, but one which had been used to create vast forests in New Zealand for the production of pulp and paper, for export as logs or chips and for timber. That emphasis did not fit his dream. After completing his degree he returned to Hawke's Bay and applied successfully for a job with the Parks and Reserves Department of the Hastings City Council. Placed in charge of the grounds of the Windsor Park Motor Camp, he saw potential there. He was able to create rock gardens and native gardens. Garry uncovered an old well and set up a water garden using limestone rock. He worked with perennials and undertook herbaceous work.

He was soon back on the local committee of the Forest and Bird Society, and also became secretary-treasurer and newsletter editor for the Hawke's Bay branch of the New Zealand Tree Crops Association, filling those positions for the next seven years. His involvement with the association created another important connection: the branch chairman was Dr Don McKenzie, pomologist at the Department of Scientific and Industrial Research

Besides making decisions, trust board members are also called on to assist with working bees in the Homestead Garden. Here Rodney Faulkner and Paul Pollock remove the standard rose 'Sander's White' to clear the ramp for wheelchair access to the House Terrace and allow for development about the urns. (Photo Garry Clapperton)

In 1988 the QEII National Trust made financial contributions towards Eastwoodhill but this ended when the Government changed the QEII's funding arrangements. Work completed through this assistance included the removal of the old shelterbelt and woodlot stumps and the smoothing and grassing down of those areas. This visually improved the public perception of the arboretum. Here 60-year-old pine stumps from along The Vista are burned on the Fountain Lawn. (Photo Garry Clapperton)

orchard in Havelock North. Dr McKenzie's research on the breeding of apples, and his association with Japanese and Chinese scientists, fascinated Garry and he spent many absorbing hours at the orchard. He commented later: "Don McKenzie expected a lot of me. He encouraged me to reach out and he built up my confidence." Dr McKenzie's enthusiasm for the Chinese people and their botanists rubbed off too. It certainly influenced Garry's decision to visit China many years later, in 1992 and again in 1994.

In 1978 he accepted a position with the Presbyterian Social Services Association as live-in house father "on site" in the evenings, at Hillsbrook Children's Home. After a month he resigned from his position at the council and served at the children's home fulltime for the next three and a half years.

Garry became aware, at the end of 1981, that the Eastwoodhill Trust Board was seeking a curator for the arboretum. Perhaps his time had arrived. He applied, and was invited to attend an interview in Gisborne. It is difficult to imagine the depth of disappointment he must have felt when his application was declined because the board considered at that time that a married

Eastwoodhill has attracted visitors from around the world. Lincoln University, the Forest Research Institute and the Plant Materials Centre have directed scientists from USA, China and Turkey to visit the arboretum. The group pictured is of forest scientists from China. The man leaning against the magnolia in the centre is Wang Shiji, Director of the Research Institute of Forestry in Beijing, and kneeling to the left of him, is Allan Wilkinson from Aokautere, who is responsible for the breeding of new poplar cultivars in New Zealand and has established three poplar trials on Eastwoodhill. (Photo Gisborne Herald)

A severe wind storm in November 1994 caused extensive damage to some areas of Eastwoodhill. In the Homestead Garden this **Quercus coccinea 'Splendens'** was toppled. After the log was cut off, the stump reverted to an upright position and both the red oak rootstock and 'Splendens' graft put out shoots, allowing propagation of this scarlet oak cultivar. Pictured are (from left) garden volunteer Betty Hair, curator's assistant Greg Papworth and trust board member Paul Pollock.
(Photo Rodney Faulkner)

couple would be more suitable for the position. He moved on to Hereworth School in Havelock North as housemaster and tutor. Meanwhile, he developed a new botanic connection: the committee chairman of the Hillsbrook Children's Home, Mr Rodney Gallen, QC, owned a 12-acre private arboretum. It was, of course, a magnet to Garry. Then Rodney's sister employed him to set up a water garden on a family property at Maraekakaho, on the south bank of the Ngaruroro River. The project involved hauling by tractor and trailer 200 tonnes of rocks to build the terraces. He planted trees, shrubs, perennials, bulbs and water plants in order to create the desired effect.

Three years had passed since he had applied for the curator's position at Eastwoodhill when he received a phone call from Terence Williams in Gisborne, informing him that it was vacant again following the departure of Kevin Boyce. This time the board did not advertise: it was offering the post to him.

"I'm afraid I'm still not married," Garry told him.

Terence Williams replied: "That's alright now. The board's view has changed."

And so began another new era for Eastwoodhill. Ten years of clean-up were behind the board. During that period there had

For several years Eastwoodhill had 'permanent loan' of the hydraulic trailer belonging to neighbours Mike & Phillida Eivers. It was invaluable for the arboretum's two-strong staff. The funds made available by the 1992 Eastland Garden Festival enabled the purchase of the arboretum's own trailer. Here Greg Papworth spreads metal on the Circus Walk, which enables Periodic Detention vehicles, in particular, access into Douglas Park throughout the winter months. The tractor was purchased with a grant from the Lottery Grants Board.
(Photo Garry Clapperton)

A recent project is the establishment of a native tree walk that will incorporate Douglas Cook's 1936 planting at Yunnan Court, natural regeneration in Cabin Park and the 1979 plantings in Gondwana Bush. From left are Lottery Grants Board representative Elizabeth Kerekere, trust board member Rodney Faulkner and curator Garry Clapperton. (Photo Gisborne Herald)

been care to preserve valuable rare plants, but virtually no new material had been planted. Now the board had employed a botanist and a plantsman with wide experience in taxonomy. Garry Clapperton was the right man, and 1985 was the right time, to begin further development of the arboretum.

Within the following 10 years the planted area at Eastwoodhill of 160 acres had "grown" to a collection covering 200 acres. Added to this, five acres of very steep country had been planted in *Cupressus lusitanica*; 18 acres were space planted, but still grazed, in three different poplar hybrid trials conducted by Landcare Research (previously the Department of Scientific and Industrial Research of Aokautere); and 15 acres of steep, eroding land had been close-planted in *Pinus radiata* under a government scheme.

*Violent winds again caused damage in November 1996 when an apparently-healthy 40 metre tall specimen of **Eucalyptus pulchella** fell, destroying the five-year-old tractor sheds. Inspection revealed major damage caused by Armillaria fungus that had infected about 95% of the trunk at ground level, with no signs of this on the outside. With limited staff numbers, the clean-up after such events greatly affect normal work at Eastwoodhill. (Photo Garry Clapperton)*

14 New Friends

The Eastwoodhill Trust Board acted positively on the idea of establishing a new Friends of Eastwoodhill Association, with a formal constitution, to raise funds and to promote the arboretum. It called a meeting in the city council chambers on April 26, 1985 in order to set up a steering committee. With typical efficiency, Bill and Terence Williams had approached Dick Richardson, a former hotelier now retired to home and garden in the city, to act as the committee's chairman. Dick was known, particularly through tennis, as an able administrator. He had been national president of the New Zealand Lawn Tennis Association and had enjoyed a long association with the Williams family through their local sponsorship of that sport.

More than 80 people gathered in the council chambers. Bill Williams, who chaired the meeting, told them that he hoped the arboretum would be promoted throughout New Zealand and the world as it deserved, as the collection was probably the finest grouping of Northern Hemisphere trees in the Southern Hemisphere. He thanked the people of Gisborne for their support,

Prior to the building of the Douglas Cook Centre, members of the Friends manned the gate from The Barn. The gatekeeper's chalet was built in the carpark in 1991 and volunteer Friends were rostered to the gate every Sunday except for winter months. An honesty box operates for other times. The notice-board was built with funds from the 1992 Eastland Garden Festival and the carpark sealed with grants from Williams & Kettle and the Gisborne District Council. The entrance drive was sealed with a grant from the A R Pierce Charitable Trust. (Photo Garry Clapperton)

Members of the Friends volunteered time and vehicles to set up stands at the Gisborne and Hawke's Bay A & P shows, Poverty Bay Horticultural Society flower shows, Mystery Creek Fieldays and Ellerslie Garden Shows to spread the word about 'The Jewel of Eastland'. (Photo Friends of Eastwoodhill)

Jenny Barker (left) and Snow Hansen in The Barn. Jenny was famous for her banksia and pine cone people and animals. Her husband Pip constructed a 'fireplace' to display silver birch logs which were for sale. Initially only one part of the double garage was used but public support resulted in both sides being used. (Photo Friends of Eastwoodhill)

but said it was now time to take stock and to extend recognition of the arboretum further afield. Allan Wallbank, then Member of Parliament for Gisborne, spoke of the debt of gratitude which New Zealand owed the Williams family for securing the arboretum as a national asset, and called for more people to join Friends of Eastwoodhill "as a duty" so that the arboretum could be enhanced for future generations. Allan Wallbank was to become active himself in spreading the gospel among leading politicians.

The board appointed Dick Richardson chairman with the approval of the meeting, and Terence Williams was to be the board's representative on the committee. The meeting elected Judith Franks as secretary and Wendy Dodgshun treasurer. It also elected to the steering committee Dawn Jefferd, Spencer Bush, Gary Quinn, Elwin (Snow) Hansen, Annie Millton and Leslie Pittar.

Even at that early stage, with the association yet to be formalised with a constitution, members went into action. The meeting had been told of space available at a leisure and pleasure exhibition that was to be staged in Gisborne's Army Hall on June 8 and 9 by the Community Arts Council. The committee immediately began planning a slide/tape presentation and, as a fund-raiser, prepared jellies and fruit preserves made from Eastwoodhill crab apples, for sale at the display. Walnuts, sweet chestnuts and maple seedlings from the arboretum would also be sold.

The first meeting of the committee of the Friends of Eastwoodhill Association took place on June 27.

Garry Clapperton, the new curator, arrived at Eastwoodhill late in August immediately after a major Ngatapa flood, with all the clear-up that entailed. But he quickly became involved with the Friends. Within two days of his arrival, in fact, he joined committee members in the first of many out-of-town promotions – an exhibition using a map, aerial and coloured photographs, botanical samples, four Chrissy Watkins watercolours and an automatic slide presentation at the Ellerslie Garden Festival, Auckland. Garry helped to man the stand, along with Bill Williams,

The Friends actively promote Eastwoodhill to dignitaries, Gisborne District Council, Motelier's Association, Visitor Information Network conference, and business groups as opportunities arise. The pavilion was used prior to the opening of the Douglas Cook Centre. Pictured in 1986 (from left) are Friends chairman Dick Richardson, Gisborne Mayor Hink Healey and Deputy Mayor, and later trust board member, Brian Crawshaw. (Photo Friends of Eastwoodhill)

Terence Williams, Jill Kindle and Chris Williams who was living in Auckland. Gary Quinn's voice had been recorded for the slide presentation, and Terence reported: "We played it continuously – so much so that Gary's voice became so soft, we had to let it cool off and regain volume from time to time!" He was delighted that they had signed up 35 new members "with some more to come".

Central to the goals of the Friends was the need to help the trust board build up the Endowment Fund for investment. It needed to be sufficiently large for the interest to finance future costs and development of the arboretum. The association established two sub-committees – publicity and increased membership; and fund-raising. All donations, membership fees and income from promotions and sales were paid, 100 per cent, to the board. The Friends claimed on the board for the relatively small amount of money it needed to run its own organisation. As an incentive, Bill Williams made yet another generous gesture, offering from the Turihaua Trust a dollar-for-dollar subsidy of up to $50,000 on money raised by the Friends. An early priority for the board was to build the income needed to provide for an assistant to Garry Clapperton. Snow Hansen has observed: "We achieved that in fairly quick time." (Michael Higgs was the first assistant appointed, followed later by Greg Papworth).

The committee decided, with the trust board's approval, to set up a shop at the arboretum. Women volunteers, assisted by handymen, converted Douglas Cook's former double garage into The Barn. The women cooked chutneys, jams and sauces for sale on open days. They made Eastwoodhill souvenirs such as wall hangings incorporating seeds and decorative cones from the arboretum. Two enthusiastic contributors were Jenny and Pip Barker of Glenroy Station, Whangara. Pip had skills as a wood-turner, producing objects of art in wood; Jenny was adept in handicraft. The Barn was to remain an Eastwoodhill institution until 1990 when it was demolished to make way for the Douglas Cook Centre. Sales were then moved to the gate-keeper's chalet.

The Friends committee showed flair and imagination both in promotion and fund-raising, and never hesitated to recruit the help of people with specific skills for particular projects. Fund-raising included a recital by international concert pianist Georgina Zellan-Smith, sponsored by the Dawn Meat Company of Hastings. It raised $1553. One rare failure was a fashion parade, which recorded a loss of $271. There were raffles and competitions of all kinds. Promotional exhibits were mounted at the local and Hawke's Bay Agricultural and Pastoral shows. Nisbet Smith, a horticultural tutor at the Tairawhiti Polytechnic, agreed to edit a regular Friends newsletter, and an important publicity tool was created when the Westpac Bank funded, with a donation of $2500, production of a coloured Eastwoodhill brochure.

At the first annual meeting on May 28, 1986 Dick Richardson was able to report an excess of income over expenditure of $9289, and the committee had been associated with a donation of $10,000 from the L.J. Renner Charitable Trust. Membership of the Friends had risen from below 100 to 573. Increased local awareness of the arboretum was stimulated by Ivan Mitchell, who began writing a series of more than 70 articles on Eastwoodhill's history in the Weekend Extra edition of the Gisborne Herald. He used the nom de plume Sambucus - the generic name for elderberry. It was an exercise involving innumerable hours of research. An important gain in that year was the granting of charitable status to the Friends, allowing tax deductions for donations.

The association linked itself with the Eastland Promotion Council to include Eastwoodhill in national and international publicity as "the Jewel of Eastland". Photographs of the arboretum were included in a New Zealand tourism display at an international conference of the Friendship Force in Hong Kong; tape cassettes of Gisborne attractions, including Eastwoodhill, were made available through New Zealand in rental campervans; and there was an Eastland carriage on a Holidaymaker train tour of North Island centres which association members helped to man.

The Friends achieved a major coup early in 1989 when Professor David Bellamy accepted their invitation to visit Eastwoodhill and give an address there. They towed a trailer out to the arboretum and created a speaking platform. The gate takings amounted to $2473. More importantly, Professor Bellamy, an international figure in conservation, became enthusiastic about Eastwoodhill and included a chapter on it in his book Moa's Ark, published in 1990. He wrote of Douglas Cook's dream to expand the collection to include as many of the trees of the Northern

In February 1989 over 600 people gathered in the Daffodil Patch to hear David Bellamy. One moment long remembered was when he asked the crowd to hold their breath. After a lengthy pause they were told to let it go. He told the crowd the trees would be saying "thank you, thank you" as a bonus quantity of carbon dioxide spread through the surrounding foliage. (Photo Friends of Eastwoodhill)

Spencer Bush (left) advocated the formation of a formally-constituted Friends of Eastwoodhill and was later arboretum supervisor for the trust board. Marion MacKay (right) began her Eastwoodhill association during Easter 1979 when she and another Massey Horticulture student made a study which they submitted to the trust board.

Hemisphere as possible, far away from radio-active fallout and acid rain. He added: "Now I have met the Friends of Eastwoodhill, I know it's going to happen..." David Bellamy is himself a member of the association: the Friends elected him an honorary life member.

The gathering momentum in recognition of Eastwoodhill continued through 1989-90 with Nisbet Smith in the chairman's seat. There were amendments to the constitution of the Friends at the 1989 annual meeting which included a rather broader view of the association's objects: "(a) To act for the general public benefit in the dissemination of knowledge of silviculture and the preservation of native and exotic flora; (b) To extend general public awareness and knowledge of native and exotic flora; (c) To promote interest in, to encourage and to assist with the maintenance and development of Eastwoodhill as an arboretum; (d) To raise funds for the Eastwoodhill Trust Board."

The Friends elected a woman to the chair for the first time in 1990 in the person of Bett Chrisp. They also elected her husband Tony as secretary. The couple had an association with Eastwoodhill dating back to 1959. They met Douglas Cook when Tony was involved with the Gisborne Jaycees on one of the Daffodil Days. Later he acted as his lawyer. During Bett's three-year term as chairman of Friends, the association prepared a handsome brochure, with help from Dunstan & Kinge in compiling photographs. It was used in a fund-raising venture for the Douglas Cook Centre and equipment needed. She wrote of the centre: "The Eastwoodhill Arboretum constitutes the most important tree collection in New Zealand in diversity and maturity. As such it represents the opportunity to establish a major focus for scientific and educational aspects of ornamental horticulture." It was a hint of new directions at the arboretum.

An idea initiated by the East Coast branch of Women's Division Federated Farmers in 1992 grew into an event of spectacular proportions. The division's president, Jan Sinton, recalled: "We started with nothing but a copy of the Taranaki Rhododendron Festival and heaps of enthusiasm." The plan was a 10-day Eastland Garden Festival involving 130 private gardens and the Eastwoodhill Arboretum. The trust board was to receive the profits. The Friends quickly became involved. Ian McIldowie served on the Garden Festival Committee which was led by Peter Franks, a beef breeder and chairman of Port Gisborne Ltd. Sponsorship, which included substantial administrative and publicity costs, was provided by Trust Bank Central whose marketing manager David Clapperton was, coincidentally, Garry's cousin.

The organisational task was enormous. The festival committee decided to incorporate as many activities as they could "so people could experience as much of Eastland as possible". Festival programmes were produced for sale. Special events included a fun debate, organised by a committee headed by Gisborne businesswoman Bronwen Holdsworth. The debaters were Gary McCormick, Maggie Barry, Kerre Woodham, Jim Hopkins, Pene Walsh and Bevan Turnpenny. The debate alone raised $2552. Young people were involved: profits of $200 from a Gisborne Boys' and Girls' high schools' Theatre Sports Show were handed by Norman Maclean and Yve Gould to Bett Chrisp. Total profits from the festival were more than $30,000.

Ian McIldowie, a viticulturist, succeeded Bett Chrisp in 1994 – a year in which the Eastwoodhill Trust Act, 1975 was amended to allow a member of the Friends to be elected to the trust board. Coincidentally, Lee Newman, secretary of the board, also became secretary of the Friends: coincidentally in the sense that, as a dedicated Eastwoodhill supporter and a member of the Friends, she simply agreed to do both jobs at once.

In the 1994-95 financial year the association raised $60,000 for the board. Ian McIldowie saw a growing source of income into the future from catering in the Douglas Cook Centre for bus tours, luncheons, meetings and seminars. He looked optimistically to the years ahead.

By December 1996 the Endowment Fund stood at $1,217,219 – providing, through the interest earned, income to sustain and further develop the arboretum. The Friends had contributed $217,462 of that total. The other sources, in addition to the original gift from the M.A. Williams Charitable Trust of $50,000 were: J.N. Williams Memorial Trust $200,000; Frimley Foundation $5000; Springhill Charitable Trust $15,000; Turihaua Charitable Trust $50,000; Turanga Charitable Trust $250,000; M.A. Williams Charitable Trust $150,000; Sundry $45,997, Surplus on sale of investments $233,760.

Many people had reason to be proud of the achievement.

15 A New Era

A new era was germinating at Eastwoodhill in the mid-1980s. Through the ensuing 10 years it blossomed in many forms – science and education, propagation, new plantings, improved access to the arboretum and better signage, greatly enhanced public knowledge about Eastwoodhill, the encouragement of people not only to enjoy the beauty of the tree garden but to understand its botanical significance. By August, 1994 Garry Clapperton was able to comment, in an address to the first New Zealand Tree Symposium of the Royal New Zealand Institute of Horticulture in Rotorua: "We are a small and largely amateur group of people who are attempting, somewhat successfully, to take Eastwoodhill from a large private tree garden, into the next century as a relevant, progressive, interpreted, educational arboretum, to assist in the process… of educating New Zealand to the benefits of trees and a care for the landscape."

Periodic Detention workers made a number of seats for the carpark picnic area utilising Japanese cedar milled from Eastwoodhill trees. Here Friends chairman Bett Chrisp (left) and curator Garry Clapperton (centre) take delivery of them from Periodic Detention warden Des Omundsen. (Photo Gisborne Herald)

Four principal elements interacted to achieve those developments – the resolve of the trust board to move forward after the long years of clean-up; the formation in 1985 of the Friends of Eastwoodhill Association with its impact on fund-raising and promotion; the appointment, for the first time, of a botanist to the position of curator; and interest from educational institutions.

The clean-up process was not complete when the PEP scheme ended, just before the appointment of Garry Clapperton. However, the board decided in 1986 to take advantage of a different scheme, operated by the Justice Department, through which people were sentenced by courts to work, periodically, on community projects. It was called "periodic detention". The PD

A passenger trailer donated by the J.& T. Hickey Charitable Trust, which also donated a portable speaker system and a wheelchair, enables those unable to easily walk Eastwoodhill to enjoy part of the arboretum. Here trust board member Don Graham sets out from the Douglas Cook Centre on a 25 minute tour of the garden and Douglas Park with visitors from the Arohaina Resource Centre. (Photo Hillary Kennedy)

gangs in Gisborne were organised by Des Omundsen, warden of the local Periodic Detention Centre, who engaged as his first site supervisor for Eastwoodhill a retired police constable, R.S. ("Blackie") Read. Blackie was well known in the district as a former amateur wrestler and surf life-saver. He combined police experience with a talent to relate to people. On the Eastwoodhill project, he drove the gangs out to Ngatapa in a Justice Department van.

He was followed as supervisor by Ruth Nepe, then Mere Edwards, who served at Eastwoodhill for five and a half years before retiring in 1994. It was a wrench for her when she left: "It has the most soothing feeling up here. It's the place to go if you have stress in your life. " And of the workers, she commented: "It gave them a sense of purpose and achievement. Some people may consider the workers rough-necks, but they have a great respect for the arboretum." Significantly, Mere retained a link with the property as a member of the Friends.

Apart from the hard slog of clean-up, PD workers have been used over the years for creative projects, most notably the formation of the H.B. Williams Walk, named after the board chairman. It was decided to construct the walk to open up the area known as The Vista. It also drew in an old fountain and a group of miniature conifers and incorporated a planting of 11 Kate Sheppard camellias (*Camellia japonica*). The plants were donated by various groups to honour the pioneer suffragette during the 1993 centennial of women's suffrage in New Zealand.

An important aspect of the trail was its role, after grading and levelling, in improving access for wheelchairs to Douglas Park. Donations from the Sir John Logan Campbell Residuary Trust, and grants from the Gisborne District Council, enabled the whole walk to be metalled, giving all-weather access to walkers as well as to wheelchairs. With the formation of the paths, the Friends were able to approach the J. and T. Hickey Charitable Trust for the funds needed for a wheelchair. The same trust donated the money to build a trailer specifically designed for the carriage of elderly or

Stephen Jones

Magnolia x soulangiana 'Rustica Rubura'
rhododendrons and common beech above The
Fountain, Douglas Park.

Friends of Eastwoodhill man the gate on a spring open day in 1986.

Azolla rubra, *the floating red water fern on Blackwater, Douglas Park.*

The Daffodil Patch, Corner Park.

John Holdsworth

Stephen Jones

Under the Patula pines on The Highway

Scarlet oaks, liquidambars and liriodendrons feature in views over Cabin Park and Douglas Park. The trees contrast well with the surrounding bare hills.

Tree labelling at Eastwoodhill is made simple by donations from two Gisborne firms and the proceeds of the 1992 Eastland Garden Festival. Karaka green Colorsteel labels, provided by B.J. Moss, hang from sharpened fibreglass pegs donated by Pultron Composites. (Photo Garry Clapperton)

disabled folk who were unable to walk the paths and enjoy the arboretum. The trailer is towed behind a small tractor.

Of the new walk, Garry Clapperton was to observe: "Naming the trail after Bill Williams is to acknowledge the enormous contribution he has made. He is totally behind Eastwoodhill. He gave it to New Zealand and regularly supports it through family donations. His whole heart is in it." Garry added that much credit for the trail must go to the PD workers – "They put their hearts and souls into it."

From as early as 1985 Terence Williams, as supervisor, had been anxious to improve access for the public through identification of tracks. He obtained cattle ear-tags from the local freezing-works. Three selected colours were used and the signs, marked with a black arrow, were fixed to trees, fence posts or gates along certain tracks. The system enabled visitors to follow a route that would return them to the carpark. (Many visitors had feared to explore too far in case they "got lost".) However, the system was limited and in 1993 a new one was developed, based on a concept used at Whakarewarewa Forest Park in Rotorua. Square steel plates covered in a coloured plastic coating, and with a white directional arrow, were fixed to half-round posts at each track junction. This project was sponsored by the funds granted from the 1992 Eastland Garden Festival.

All of the funds raised by the festival were donated to Eastwoodhill. They enabled progress on a number of projects, including labelling, signing and notice boards, seating and shelters. With the new system for identification of tracks in place, even unaccompanied children were now able to explore the arboretum with confidence.

Tree labelling had begun, in a limited way, in the days of Dan Weatherall with painted flat steel labels fixed to trees using nails. In 1985 an innovative concept from Terence Williams was developed. He used a sharpened fibre glass skewer which was driven into the trunk of the tree, two metres above the ground. A large, easy to read Colorsteel label was hung on the peg. The labels could move in the wind and the pegs would not affect a saw. The problem of the trunk growing over a label was solved. Small aluminium labels were used on younger trees and on other trees off the main paths.

Interpretative printed material, enabling people to understand aspects of the tree collection, was another aim, along with educational material for the thousand primary and intermediate school children visiting in groups each year. There was recognition of the need "to assist in the moulding of young minds regarding tree consciousness and all its associated values". The announced aim for primary school education was to produce simple, precise exercises "that can open kids' eyes to some aspects of the trees, ecosystems and lifecycles".

The year 1985 had brought a new initiative in "mapping" the trees of the arboretum, and created a link with Palmerston North's Massey University, an institution with strong horticultural and agricultural associations. Marion MacKay, a horticultural lecturer and research scientist at the university, tackled the task, spread over three years. She relied heavily on Bob Berry's catalogue as revised by him in 1982. One student had done some preliminary work as a holiday project. Marion set out to convert the whole Bob Berry catalogue into the form of maps – the first steps towards the creation of a computerised data base.

It was a painstaking process. During numerous visits to Eastwoodhill she trod every square metre of the arboretum. By the winter of 1988 her draft maps covered 36 sheets which recorded all physical features – including trees, tracks, fencelines and waterways. A major challenge was to identify rare trees which Bob Berry had recorded but had not been able to name. That meant hours of work studying botanical dictionaries, checking such clues as leaf, flower or seed characteristics. Garry Clapperton was involved in data entry and field checking as well as reference to Douglas Cook's notes of purchases and plantings. The noted botanist Bill Sykes assisted in solving particular puzzles, as did other knowledgeable plantsmen who visited the arboretum.

Marion MacKay's work at Eastwoodhill led, in 1987, to a visit by Professor David Chalmers, professor of horticultural science at Massey. Like other botanists before him, he was enthralled. From the time the trust board had been established, the chairman and members were conscious of the scientific and

Marion MacKay of Massey University was a member of the advisory committee when she started a mapping project as her own contribution towards Eastwoodhill. This was many months before her department head suggested she use the project towards her Ph.D., which she was awarded in 1997. (Photo Massey University)

educational value of the plant collection and looked forward to the development of research and educational facilities at some time in the future. In Bill Williams' words, the professor "prodded us along" by suggesting that the board build "an Eastwoodhill Centre for Ornamental Plant Studies". Professor Chalmers considered that such a centre would "develop a scientific focus and attract other groups to an arboretum that already has an international reputation".

The beginning of visits by arboricultural students from the Waikato Polytechnic to practise tree surgery techniques added weight to the proposal. These visits benefited both students, through practical tuition, and the arboretum by having badly needed tree surgery carried out. Regular visits by Massey University began the following year with similarly beneficial results. The educational value to students included time spent with the curator learning tree identification and other aspects of the collection.

The first estimate of $900,000 for an Eastwoodhill educational centre was something of a shock for the board. Undaunted, however, it approved a plan for a Lockwood building for the more digestible cost of $300,000. A $100,000 grant towards the project from the L.J. Renner Estate was heartening, to say the least. Bill Williams matched that grant from various family trusts and, in addition, offered to fund any shortfall, interest-free for 12 months. A three-stage construction plan was envisaged, with the first two stages concentrating on plant studies and the third to be a separate accommodation block at a later stage.

It was with the encouragement of Prof. David Chalmers of Massey University that the trust board pursued the building of a plants study centre. Outside the newly-completed Douglas Cook Centre in 1991 are from left, Bett Chrisp, Don Graham, Spencer Bush, Terence Williams, H.B. Williams, David Chalmers, Marion MacKay, Garry Clapperton, Anne Berry and Bob Berry.

One of three seating shelters built in Douglas Park and Glen Douglas with funds from the 1992 Eastland Garden Festival.

The board decided to call the facility the Douglas Cook Centre for Education.

Major donors who helped to make the project possible are listed on a board in the centre: the J.N. Williams Memorial Trust, Mangatawa Charitable Trust, Springhill Charitable Trust, Turanga Trust Board, Frimley Foundation, L.J. Renner Estate, Stanley Smith Horticultural Trust, New Zealand Lotteries Board, Trust Bank Central, and the Sir John Logan Campbell Residuary Trust.

Members of the Friends and other supporters were always close by for chores that would help to keep costs down. Dr Maurice McLachlan donated five tables, handmade from Eastwoodhill decorative timbers. Side benches for the laboratory were made from Eastwoodhill red oak. With book cabinets in place, the valuable Douglas Cook collection of 845 botanical books and documents was returned to Eastwoodhill from the H.B. Williams Memorial Library where it had been held in safekeeping for 20 years.

The official opening of the centre on October 25, 1992 was a focal point of the Eastland Garden Festival. It was wet – a day for umbrellas and, at the end of the ceremonies, for hauling bogged vehicles off the slippery Parking Green by tractor. But spirits were high. The Minister of Conservation, the Hon. Denis Marshall, spoke of Douglas Cook's "magnificent obsession" and said: "His legacy lives on, largely thanks to the Williams family who intervened to make sure that the arboretum became a national treasure." Marion MacKay pointed out that science and education had always been part of Eastwoodhill philosophy – "Douglas Cook was keen to share his knowledge." Professor Chalmers, who was now attached to the Charles Sturt University of Australia, said of Eastwoodhill: "It has inspired me and I have seen it inspire my colleagues and my students. Douglas Cook's vision is on course." Bill Williams described the centre as a lasting memorial to Douglas Cook. Official guests included Mrs Della Newman, United States Ambassador to New Zealand. There were blessings by the Rev. Huatahi Niania and the Rev. Pamela Tankersley.

To cut the ribbon, the board had invited Maggie Barry, national broadcaster and presenter of the television programme, Palmers Garden Show. She called Eastwoodhill "the jewel in the crown of the Garden of Eden" and added: "We should all feel proud of one man's vision and what has been created by that vision and the work of a group of local people." Douglas Cook would have been proud, not only because of the event itself, but also because of the presence there of three family members. His adopted son, Sholto, unveiled a wooden sign carrying the centre's

Television personality Maggie Barry addresses an attentive crowd, braving the rain in October 1992, prior to cutting the ribbon and opening the Douglas Cook Centre. (Photo Brian Crawshaw)

name, and the guests included Jan Potter (daughter of Douglas Cook's sister, Sheila Wily) and Susan Perry (grand-daughter of Sheila).

The invitation to Maggie Barry to cut the ribbon proved to be a most rewarding choice. She was enchanted by what she saw and heard at the arboretum and decided to return in the autumn of 1993 with a television crew to tape a segment for her Palmers programme. Earlier, thanks largely to the enthusiasm and energy of the Friends, there had been growing media interest in Eastwoodhill with various newspaper and magazine articles, coverage in the National Radio programme Roundabout, and television exposure in Weekend and Living Earth. These created valuable awareness but not a dramatic effect on actual visitor numbers. Maggie Barry's programme, however, was telecast at peak hour and beamed specifically to garden lovers who regularly watched the show. Its impact was immediate: there was a deluge of mail; people flew from Auckland and Wellington and hired cars to drive to the arboretum; visitor numbers leapt by 70 per cent and gate takings for that year were a record $9200. Many thousands of television viewers still recall a sequence in which Maggie Barry and Garry Clapperton lay side by side on the ground, looking skyward to study the fascinating patterns created by the leaves and branches of the Patula pines.

In 1987, the year during which the idea of the education centre was developing, the trust board adopted its management plan and also decided to create a Plant Management Committee of invited plantsmen. Professor Chalmers was the first convenor and was joined by a group of eminently knowledgeable people – Peter Cave, Ian McKean, Ron Gordon, Michael Hudson, Bill Sykes, Alan Jellyman, Gordon Collier, Bob Berry and Marion MacKay. The committee also included board members, the curator and supervisor "in attendance". Alan Jellyman later became convenor when Professor Chalmers moved to Australia.

While clean-up was continuing in the mid-1980s, the board was already looking both to expansion of the tree collection and

"Arboretum Distinguished for Merit by the International Dendrology Society" reads the brass plaque, the first such award presented by the IDS. The concept had come from Hugh Johnson, known as Tradescant to Royal Horticultural Society members, out of a desire to encourage gardens such as La Mortola in Italy. (Photo Gisborne Herald)

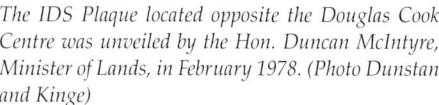

The IDS Plaque located opposite the Douglas Cook Centre was unveiled by the Hon. Duncan McIntyre, Minister of Lands, in February 1978. (Photo Dunstan and Kinge)

to ways of safeguarding precious species. The key to those developments was a network of dedicated propagators. A plantsman early on the scene was Peter Cave who visited the arboretum in 1985 to discuss the preservation of rare and endangered species. A noted nurseryman in the Waikato, Peter is the son of Harry Cave who, though best known as a former New Zealand cricket captain, was a horticulturist himself with a particular interest in camellias. Other nurseries became involved in the exchange of plant material, among them the Top Trees nursery in Clive, Hawke's Bay; Eric Appleton of Wakefield, Nelson; John and Christine Nicholls of Tauranga. Paul Pollock, who succeeded Bob Berry as the horticultural society representative on the trust board, is a commercial propagator at his Highgate property at Muriwai. He became active in propagating apple and hawthorn species from Eastwoodhill.

As the Eastwoodhill association with nurseries expanded, the curator had an increasing resource to draw upon, with different propagators having particular skills to suit a need. Allan Jellyman of New Plymouth has provided since 1988 a collection of Nepalese plants for the arboretum. In its own role of contributing material, Eastwoodhill distributed, over the space of six years, more than 60,000 acorns of its scarlet oaks to three major New Zealand nurseries. A classic case of the propagation network in action came with the loss, during galeforce winds in 1988, of one of the arboretum's unique trees - a 15-metre blue conifer, *Abies concolor*

'Candicans'. It had been imported by Douglas Cook in 1949 and he treasured it. When the winds struck, Garry Clapperton had been in the process of building a retaining wall to stabilise the tree and give it room to root. The distraught curator sent scions to three different nurseries for grafting. He received two grafts back; one survived and restored the species to the Eastwoodhill collection.

Individual members of the International Dendrology Society became a major force in propagation. Douglas Cook had purchased more than 90 per cent of all his plant material. By 1994, IDS members were donating 90 per cent of all the new plant material being used at Eastwoodhill. The horticultural successes of the board in the 1984/94 decade can be measured in size and quality. The planted area developed in that time from 160 acres to almost 200; plant varieties increased from 340 genera to over 570 genera of woody species. Success can be measured in human terms, too – "out there in our community," Garry Clapperton observed, "are some very eager individuals who have made the ornamental horticulture scene in New Zealand so much richer."

The IDS has had a long association with the arboretum and with Douglas Cook, its first New Zealand member. In the heart of Eastwoodhill there stands a bronze plaque set into a rock. Presented in 1977, it was the first award by the society, anywhere in the world, to "a collection of outstanding merit". The idea of forming a society to bring together dendrologists from throughout the world was conceived in Belgium in May, 1952 and when the first official meeting was held in September that year there were 74 members from five countries. By 1994 there were 1200 members from 50 countries. Another link with Eastwoodhill has been a personal one through Anne Berry of Hackfalls (the former Lady Anne Palmer) who served for nine years as IDS tours chairman and for nearly five years as chairman of the society.

The story of Bob and Anne Berry of Hackfalls is a classic one in terms of the bonds created by dendrology. Before they met, and many years before they wed in 1990, they were developing their own garden oases on opposite sides of the world – Bob at Tiniroto, Anne at Rosemoor in North Devon. Anne, who had New Zealand family connections, took an opportunity during a visit to Gisborne in 1970 to see Eastwoodhill. She recalled: "Despite its then run-down condition it was to me a very impressive collection, at that time managed single-handedly by Bill Crooks." When she became tours chairman she placed New Zealand on her list. The idea among the IDS Committee of awarding plaques to plant collections of outstanding merit arose at the time Anne was planning a New Zealand tour for 1977. She nominated Eastwoodhill.

Lady Anne Palmer, past chairman of the International Dendrology Society, led society tours from Britain to New Zealand in 1977 and 1990. On her marriage to Tiniroto farmer Bob Berry in 1990, the society gave them, as a wedding gift, £1,000 for Eastwoodhill which Anne requested be used to purchase modern botanical books for the library in the Douglas Cook Centre. Here she is pictured with her husband Bob Berry (left) and H.B. Williams near the IDS plaque.

Anne was able to assist Eastwoodhill in another way by becoming involved in a successful application to the Stanley Smith Horticultural Trust for funds to install a new water supply. A second grant was obtained from the trust for the establishment of a herbarium in the Douglas Cook Centre. It enabled the purchase of drying presses, cabinets, microscopes and benches in order to build a collection of dried plant material. Sally Dodgshun took up the challenge in 1994 of organising the herbarium into an active unit and the project received a significant boost in 1996 with a grant from the Lottery Grants Board.

Looking back on the progress of Eastwoodhill over the 1980s and 90s, Anne observed: "None of this would have happened but for the generous and magnanimous action by H.B. (Bill) Williams in purchasing the property, gifting it to New Zealand and setting up an endowment for its continued maintenance. Since its foundation additional sums have been added making a total of over a million dollars. Most of this has been donated by the Williams family trusts and without Bill Williams' unfailing support there is little doubt that Eastwoodhill would have ceased to exist as the arboretum such as we know today. More recently an important contribution has been made by the curator, Garry Clapperton, who since his appointment has made a remarkable impression on the arboretum both in its maintenance and expansion; also in its increased and increasing publicity and fame. Were he alive today, Douglas Cook might well be proud of his achievement, and grateful to those who have laboured so long to justify the value of his creation."

Sholto Douglas Cook, who now lives in Whakatane, is known to his family and friends as Doug. Here he unveils the sign commemorating his father at the opening of the Douglas Cook Centre for Education in 1992.

Epilogue

H.B. Williams speaking at the opening of the Douglas Cook Centre. In the background are (from left) United States Ambassador Della Newman and television personality Maggie Barry.

I was asked to assist in the preservation of Eastwoodhill because it was considered that if the property was offered for sale there was a real risk much of its uniqueness might be lost. In addition there would certainly be no opportunity to maintain and enlarge the collection particularly if the purchasers were neighbouring farmers.

Following the receipt of various reports on the value of this asset for future development and as an educational resource, I realised that should I purchase the property there would then arise the problem of preserving it for future generations. It was then decided to try to form a charitable trust. This was done after much negotiation.

Since then I have been able to assist the trust with substantial donations from many of my family charitable trusts with the aim of having an endowment of sufficient size to cover the general running costs. My efforts have been well supported by some other trusts, local and outside this district.

In those early years the task of preserving Eastwoodhill seemed so immense. We had no way of predicting the tremendous support that would come from various parts of the community, both within the Gisborne area and from other parts of the country. The value of donations in money and kind for various projects is outstanding and more has been exceeded by the voluntary efforts of enthusiasts, particularly the Friends. Without the personal commitment on the part of so many, the trust board would never have been able to oversee the progress that has been made in the protection of this remarkable asset.

The magnitude of the progress that has been and continues to be made at Eastwoodhill is satisfying to me and gives reassurance that the dream of W. Douglas Cook is a reality and will continue as such.

Eastwoodhill was the product of a labour of love on the part of one man and his helper and it is especially pleasing to see that love being furthered by the efforts of so many today.

Eastwoodhill was set aside by an Act of Parliament for scientific and educational purposes and for enjoyment by the people of New Zealand. In that regard it fulfils Douglas Cook's original intentions as regards its future. It is satisfying to me to see it being used in these ways and so magnificently supported as well.

I hope the beauty of the arboretum and its purpose will inspire many others to subscribe to the endowment in particular and to many projects which continually arise.

H.B. Williams

Index

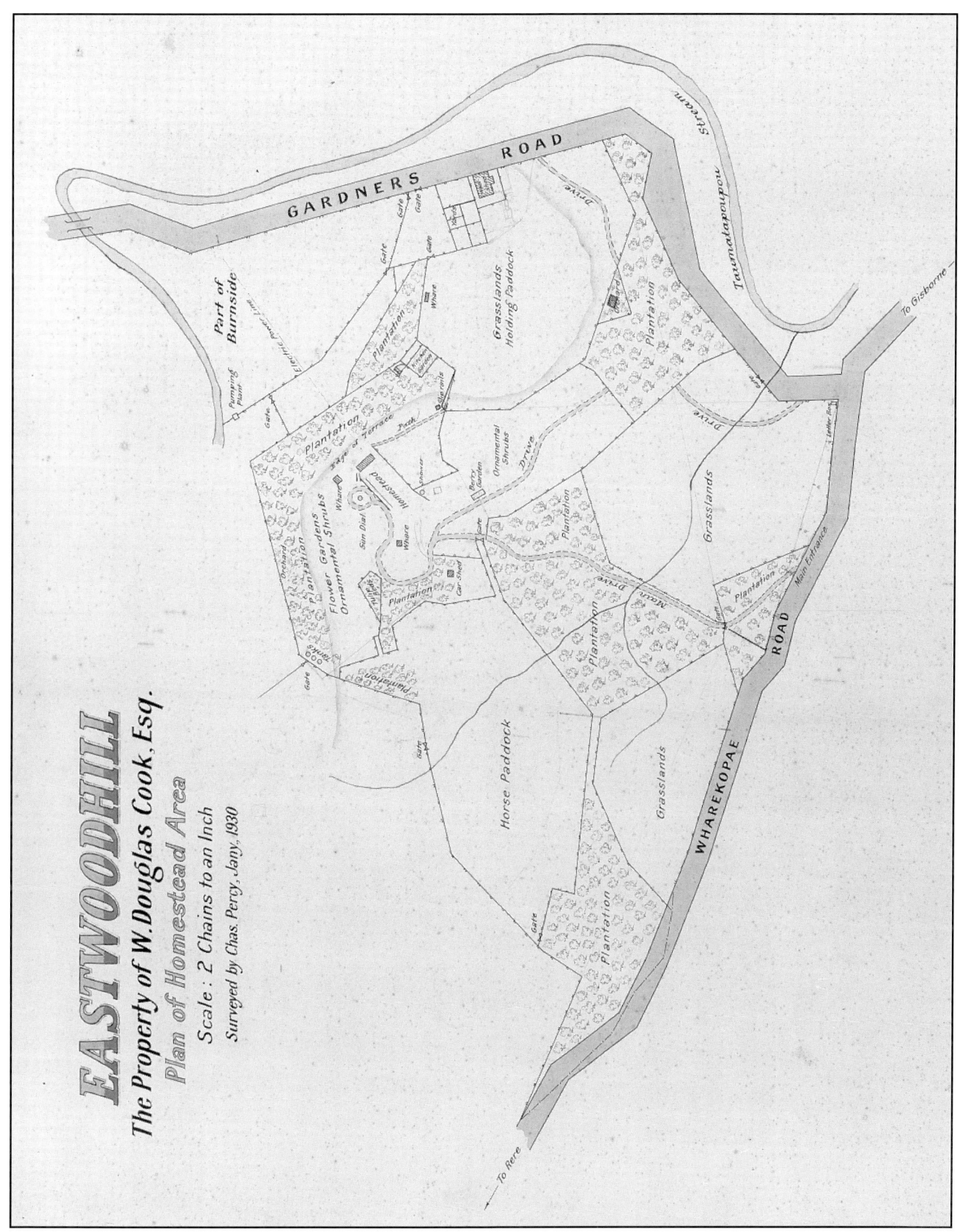

EASTWOODHILL

The Property of W. Douglas Cook, Esq.

Plan of Homestead Area

Scale : 2 Chains to an Inch
Surveyed by Chas. Percy, Jany, 1930

GARDNERS ROAD

Part of Burnside

Pumping Plant

Gate

Electric Power Line

Gate

Gate

Gate

Gate

Plantation

Whare

Plantation

Top of Terrace

Path

Plantation

Flower Gardens
Ornamental Shrubs

Whare

Sun Dial

Whare

Orchard

Plantation

Gate

Plantation

Gate
Path

Plantation

Car Shed

Plantation

Homestead

Berry Garden

Gate

Ornamental Shrubs

Drive

Drive

Grasslands
Holding Paddock

Plantation

Plantation

Plantation

Stream

Taumataipouipou

To Gisborne

Plantation

Grasslands

Main Drive

Plantation

Main Entrance

WHAREKOPAE ROAD

Grasslands

Horse Paddock

Plantation

Gate

To Rere

125

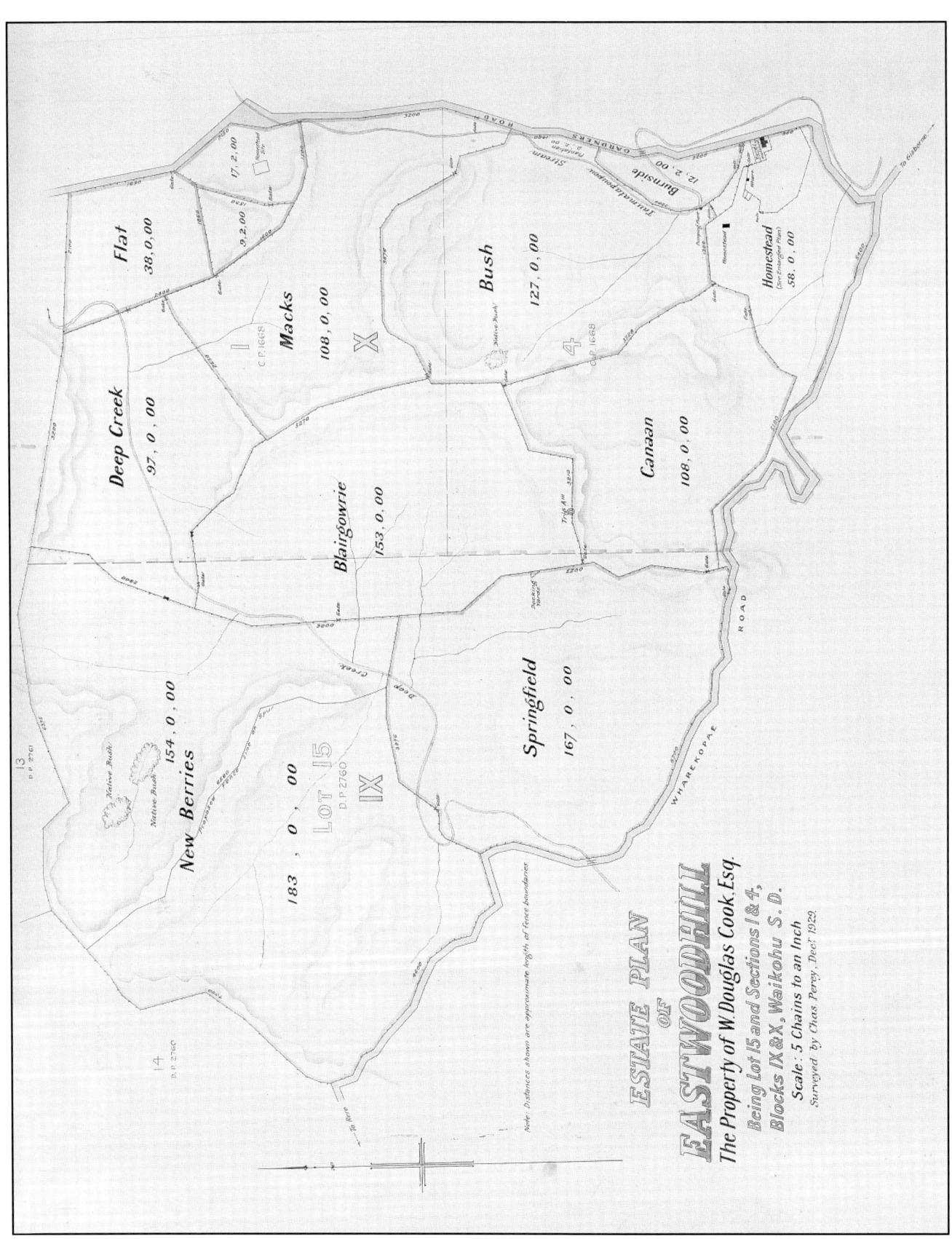

ESTATE PLAN
OF
EASTWOODHILL

The Property of W. Douglas Cook, Esq.

Being Lot 15 and Sections 1 & 4,
Blocks IX & X, Waikohu S.D.

Scale: 5 Chains to an Inch

Surveyed by Chas. Percy, Dec.r 1929

Note: Distances shown are approximate length of fence boundaries

Flat
38, 0, 00

17, 2, 00

9, 2, 00

Macks
108, 0, 00
C.P. 1668
X

Deep Creek
97, 0, 00

Blairgowrie
153, 0, 00

New Berries
154, 0, 00

Native Bush

LOT 15
D.P. 2760
IX

183, 0, 00

Springfield
167, 0, 00

Bush
127, 0, 00
C.P. 1668
4

Canaan
108, 0, 00

Trig A.W.

Burnside
12, 2, 00

Homestead
(See Enlarged Plan)
58, 0, 00

GARDNERS   ROAD

To Gisborne

WHAREKOPAE   ROAD

To Rere

13
D.P. 2761

14
D.P. 2760

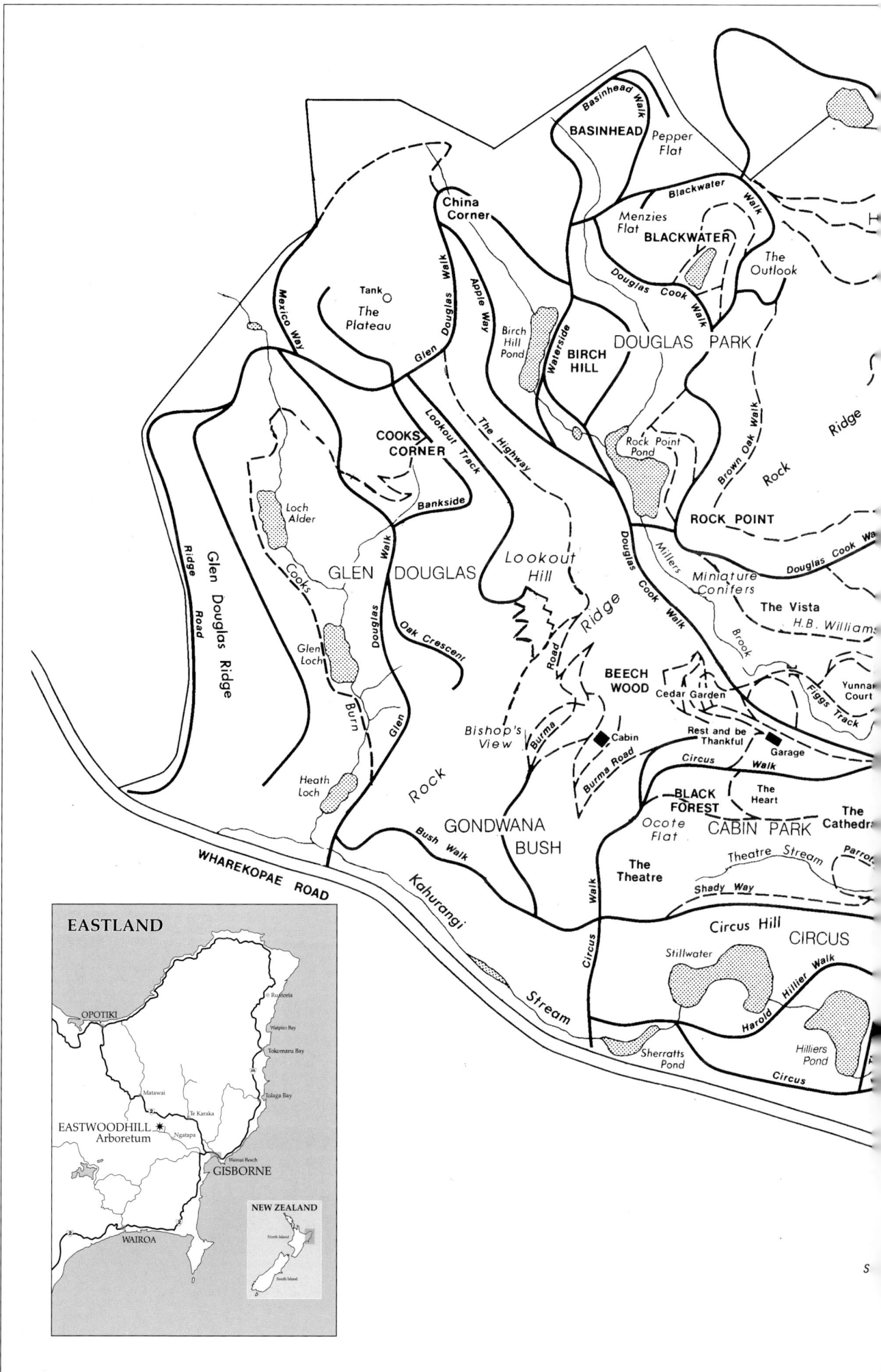

BASINHEAD
Basinhead Walk
Pepper Flat
China Corner
Blackwater Walk
Menzies Flat
BLACKWATER
The Outlook
Douglas Cook Walk
Mexico Way
Tank
The Plateau
Glen Douglas Walk
Apple Way
Birch Hill Pond
Waterside
BIRCH HILL
DOUGLAS PARK
Rock Point Pond
Brown Oak Walk
Rock Ridge
ROCK POINT
Douglas Cook Walk
COOKS CORNER
Lookout Track
The Highway
Bankside
Loch Alder
Cooks Walk
GLEN DOUGLAS
Lookout Hill
Ridge
Douglas Cook Walk
Millers Brook
Miniature Conifers
The Vista
H.B. Williams
Glen Douglas Road
Douglas Walk
Oak Crescent
Road
BEECH WOOD
Cedar Garden
Figgs Track
Yunnan Court
Glen Loch
Glen Burn
Bishop's View
Burma
Cabin
Rest and be Thankful
Garage
Circus Walk
Glen Douglas Ridge
Heath Loch
Rock
Burma Road
BLACK FOREST
The Heart
The Cathedral
Ocote Flat
CABIN PARK
Bush Walk
GONDWANA BUSH
The Theatre
Theatre Stream
Parror
Shady Way
WHAREKOPAE ROAD
Kahurangi
Circus Walk
Circus Hill
CIRCUS
Stillwater
Hillier Walk
Stream
Harold
Sherratts Pond
Hilliers Pond
Circus

EASTLAND

Ruatoria
OPOTIKI
Waipiro Bay
Tokomaru Bay
Matawai
Te Karaka
Tolaga Bay
EASTWOODHILL
Arboretum
Ngatapa
Wainui Beach
GISBORNE

NEW ZEALAND
North Island
South Island

WAIROA

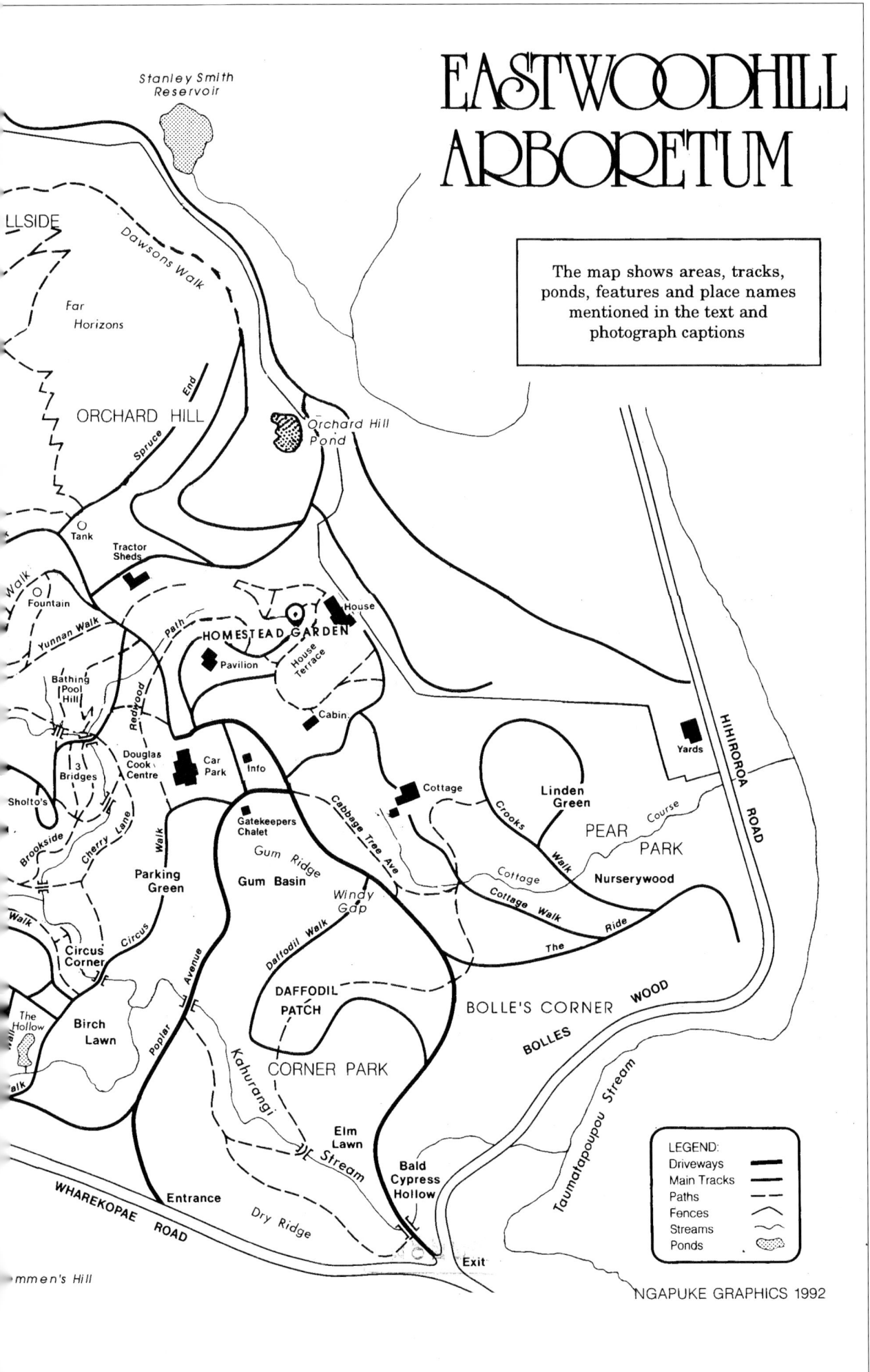

EASTWOODHILL ARBORETUM

The map shows areas, tracks, ponds, features and place names mentioned in the text and photograph captions

Stanley Smith Reservoir

LLSIDE

Dawsons Walk

Far Horizons

End

ORCHARD HILL

Spruce

Orchard Hill Pond

Tank

Tractor Sheds

Walk

Fountain

Yunnan Walk

Path

HOMESTEAD GARDEN

House

Pavilion

House Terrace

Bathing Pool Hill

Redwood

Cabin

Yards

3 Bridges

Douglas Cook Centre

Car Park

Info

Cottage

Linden Green

Sholto's

Brookside

Cherry Lane

Walk

Gatekeepers Chalet

Cabbage Tree Ave

Crooks

Walk

Course

PEAR PARK

Gum Ridge

Gum Basin

Cottage

Nurserywood

Parking Green

Windy Gap

Cottage Walk

The Ride

Walk

Circus

Circus Corner

Daffodil Walk

Avenue

BOLLE'S CORNER WOOD

The Hollow

Birch Lawn

Poplar

DAFFODIL PATCH

BOLLES

alk

Kahurangi

CORNER PARK

Taumatapoupou Stream

Elm Lawn

Stream

Bald Cypress Hollow

WHAREKOPAE ROAD

Entrance

Dry Ridge

HIHIROROA ROAD

mmen's Hill

Exit

LEGEND:
Driveways
Main Tracks
Paths
Fences
Streams
Ponds

NGAPUKE GRAPHICS 1992